T0275884

Participatory Health Through Social Media

Participatory Health Through Social Media

Edited by

Shabbir Syed-Abdul

Elia Gabarron

Annie Y.S. Lau

AMSTERDAM • BOSTON • HEIDELBERG • LONDON
NEW YORK • OXFORD • PARIS • SAN DIEGO
SAN FRANCISCO • SINGAPORE • SYDNEY • TOKYO

Academic Press is an imprint of Elsevier

Academic Press is an imprint of Elsevier
125 London Wall, London EC2Y 5AS, UK
525 B Street, Suite 1800, San Diego, CA 92101-4495, USA
50 Hampshire Street, 5th Floor, Cambridge, MA 02139, USA
The Boulevard, Langford Lane, Kidlington, Oxford OX5 1GB, UK

Notices

Knowledge and best practice in this field are constantly changing. As new research and experience broaden our
understanding, changes in research methods or professional practices, may become necessary.

Practitioners and researchers must always rely on their own experience and knowledge in evaluating and using
any information or methods described herein. In using such information or methods they should be mindful of
their own safety and the safety of others, including parties for whom they have a professional responsibility.

To the fullest extent of the law, neither the Publisher nor the authors, contributors, or editors, assume any
liability for any injury and/or damage to persons or property as a matter of products liability, negligence or
otherwise, or from any use or operation of any methods, products, instructions, or ideas contained in the
material herein.

Library of Congress Cataloging-in-Publication Data
A catalog record for this book is available from the Library of Congress

British Library Cataloguing-in-Publication Data
A catalogue record for this book is available from the British Library

ISBN: 978-0-12-809269-9

For Information on all Academic Press publications
visit our website at http://www.elsevier.com

**Working together
to grow libraries in
developing countries**

www.elsevier.com • www.bookaid.org

Publisher: Mica Haley
Acquisition Editor: Rafael E. Teixeira
Editorial Project Manager: Ana Claudia A. Garcia
Production Project Manager: Anusha Sambamoorthy
Cover Designer: MPS

Typeset by MPS Limited, Chennai, India

CONTENTS

S. Syed-Abdul, E. Gabarron, and A.Y.S. Lau

LIST OF CONTRIBUTORS

S.A. Adams
Tilburg University, Tilburg, The Netherlands

O.H. Ahmed
Bournemouth University, Bournemouth, United Kingdom; Poole
Hospital NHS Foundation Trust, Poole, Dorset, United Kingdom;
The Football Association, London, United Kingdom

S. Atique
Taipei Medical University, Taipei, Taiwan

P.D. Bamidis
Aristotle University of Thessaloniki, Thessaloniki, Greece

C.S. Bond
Bournemouth University, Bournemouth, United Kingdom

K. Denecke
Bern University of Applied Sciences, Bern, Switzerland

L. Fernández-Luque
Hamad Bin Khalifa University, Qatar Foundation, Doha, Qatar

E. Gabarron
University Hospital of North Norway, Tromsø, Norway; The Arctic
University of Norway, Tromsø, Norway

S. Hors-Fraile
University of Seville, Seville, Spain; Salumedia Tecnologías, Seville,
Spain

M. Househ
King Saud Bin Abdulaziz University for Health Sciences, Riyadh,
Saudi Arabia

E. Konstantinidis
Aristotle University of Thessaloniki, Thessaloniki, Greece

S. Konstantinidis
The University of Nottingham, Nottingham, UK

L. Laranjo
Macquarie University, Sydney, Australia

A.Y.S. Lau
Macquarie University, Sydney, NSW, Australia

A. Leis
Universitat Pompeu Fabra (UPF), Barcelona, Spain

M.A. Mayer
Universitat Pompeu Fabra (UPF), Barcelona, Spain

M. Merolli
The University of Melbourne, Melbourne, VIC, Australia

O. Rivera
University of Seville, Seville, Spain

S. Syed-Abdul
Taipei Medical University, Taipei, Taiwan

Dr. S.A. Adams is an Associate Professor of eHealth Governance and Regulation at the Tilburg Institute for Law, Technology and Society in Tilburg, the Netherlands. Her research focuses on the normativities of using Information and Communication Technologies (ICT) in healthcare and for health-related purposes. She has been an advisor to various government bodies and health institutions in the Netherlands and has personally conducted or supervised research on various types of health ICT during the last 15 years. She has also maintained a personal research line on social media and health since 2007. She is specialized in qualitative research methods and is currently experimenting with innovative methods for better understanding the use of mobile technologies for health-related purposes. She is an active member of the American Medical Informatics Association (AMIA) and is a multiple-term appointee to the AMIA Ethics Committee.

Dr. O.H. Ahmed is a Physiotherapist based in Dorset, England. He completed his physiotherapy training at the University of Nottingham, England, where he graduated with a BSc (Hons) in Physiotherapy, and went on to undertake a Postgraduate Diploma in Sports Physiotherapy at the University of Otago, New Zealand, in 2007. He completed his PhD in 2013 at the University of Otago, where he researched into the use of social media to assist return-to-play following concussion. He currently works as a clinician in the Department of Physiotherapy at Poole Hospital NHS Foundation Trust, and with elite disability football squads for the Football Association in England. He also holds an academic role as a lecturer in the Faculty of Health and Social Sciences at the Bournemouth University. He has published widely in eHealth and mHealth, and has a keen interest in using social media and online technologies to assist physiotherapy and sports rehabilitation.

Dr. S. Atique is a PhD scholar at Graduate Institute of Biomedical Informatics, Taipei Medical University, Taiwan, since 2013. His key research interests include disease surveillance especially infectious diseases, health information systems, electronic health records, nursing informatics, and role of social media in changing healthcare landscape. He is looking forward to optimize healthcare delivery system via application of Information and Communication Technologies (ICT). Currently, he is working on his thesis via an exchange program at the Institute of Public Health, Heidelberg University, Germany, since February 2016. Prior to starting his PhD program, he received his master degree in "Health Informatics" from COMSATS Institute of Information Technology, Islamabad, Pakistan. During his master degree, he also participated in DAAD-funded summer school on "Global Health Challenges in 21st century." He is also a registered Pharmacist by Punjab Pharmacy council, Pakistan, with a degree in Doctor of Pharmacy (Pharm-D) from Government College University, Faisalabad, Pakistan.

Prof. P.D. Bamidis received the Diploma degree in Physics from the Aristotle University of Thessaloniki (AUTH), Thessaloniki, Greece, in 1990, the MSc (with distinction) degree in Medical Physics from the University of Surrey, Guildford, United Kingdom, in 1992, and the PhD degree in Bioelectromagnetism and Functional Brain Analysis and Imaging from the Open University, Milton Keynes, United Kingdom, in 1996. He is currently an Associate Professor in Medical Education Informatics within the Laboratory of Medical Physics, Medical School, AUTH. He has been the coordinator of large European projects (www.longlastingmemories.eu, www.meducator.net; www.epblnet.eu, www.smokefreebrain, www.childrenhealth.eu) as well as the principal investigator for a number of national and international funded projects (more than 40 in total). His research interests are within assistive technologies (silverscience, silvergaming, mobile health, decision support, avatars), technology-enhanced learning in Medical Education (web2.0, semantic web, serious games, virtual patients, PBL, learning analytics) and Affective and Physiological Computing and HCI, (bio)medical informatics with emphasis on neurophysiological sensing and health information management (open health big data), and Affective Neurosciences. In 2009 he was awarded the Prize of the AUTH Research Committee for the Best Track Record in funded research projects among AUTH young academic staff. He has been the Chairman/Organizer of seven international conferences (iSHIMR2001, iSHIMR2005, MEDICON2010, GASMA2010, SAN2011, MEI2012, and MEI2015) and the Conference Producer of the Medical Education Informatics Conference and Spring/Summer School Series. He is a member of the Advisory Board for the Open Knowledge Foundation (OKFN), a founding member of OKFN Greek chapter, and the President of the Greek Biomedical Technology Society. In 2009, he was awarded the Prize of the AUTH Research Committee for the Best Track Record in funded research projects among AUTH young academic staff. In 2015 he was awarded the title of the Honorary Professor of Karaganda State Medical University, Kazakhstan, as well as the Pospelov Medal for his contributions to Medical Education development by the same University.

Dr. C.S. Bond is a nurse, and gained her Doctorate from the University of Bristol in the United Kingdom. She is currently a Principal Academic Digital Health in the Faculty of Health and Social Sciences at the Bournemouth University (United Kingdom). Her academic and research interests are centered around Patient 2.0, Health 2.0, and Participatory Healthcare, especially the role the Internet in supporting people living with long-term conditions. Alongside this, she is interested in developing the methodology and associated research ethics underpinning the use of online resources in health research. As well as being a registered nurse, she is a Chartered Fellow of the British Computer Society, the UK representative on the International Medical Informatics Association (IMIA) Nursing Informatics group, and a member of the IMIA Social Media working group. She is also an ehealth section editor for the *Journal of Innovation in Health Informatics*.

Dr. K. Denecke is a Professor of medical informatics at the Bern University of Applied Sciences. Her main research interests are medical language processing, information extraction, sentiment analysis, and text classification. From 2013 to 2015 she was a scientific director of the Digital Patient Modeling group at ICCAS at the Medical Faculty of the University of Leipzig. She holds a Diploma in Computer Science and was awarded a Doctoral degree in Computer Science by the Technical University of Braunschweig. Before she joined the University of Leipzig, she worked as a researcher at L3S research center in the field of Web Science and as a software architect at an IT company providing software for hospitals. She coordinated several research projects, among others the EU funded project M-Eco: Medical Ecosystem on event-driven surveillance of infectious diseases and was involved in the EU Projects Terence and LivingKnowledge. From 2014 to 2015, she worked as consultant for the Helmholtz Centre for Infection Research within the SORMAS project where a surveillance and outbreak management system for Ebola has been developed.

Dr. L. Fernández-Luque is at Qatar Computing Research Institute-Qatar Foundation, where he works on social computing for health. He graduated his PhD at the University of Tromso (Norway) and studied computer engineering at the University of Sevilla (Spain). He has been involved in eHealth and Medical Informatics research for over 10 years, working both in industry and academia in institutions such as Norut, Polytechnic University of Valencia (Spain), Harvard Medical School (United States), and the digital health startup Salumedia (Spain). In addition to editing several books about ePatients, he has more than 60 peer-reviewed publications, 30 of them indexed in PubMed and cited over 800 times in the last 5 years (according to Google Scholar). He has been working in the last 7 years in the field of Health Social Media, becoming one of the leading experts in this emerging field. He is currently the Chairman of the Social Media Working Group of the International Medical Informatics Association.

E. Gabarron received her degree in Psychology at the Autonomous University of Barcelona, Spain, in 1997. She developed her career in clinical research, field in which she has an extensive experience. And in 2012 she moved to Norway, where she works as Research Fellow at the Norwegian Centre for eHealth Research, at the University Hospital of North Norway. Since 2013, she is also collaborating as Professor for both, the degree of Psychology and the Master degree in Social Media and Health at the Open University of Catalonia, in Spain. Currently, she is also finishing her PhD in Health Sciences at the Arctic University of Norway. Her PhD research project is on the use of social media for health promotion.

S. Hors-Fraile is a researcher at the University of Seville, Spain, and Business Development Advisor at Salumedia, an SME who develops digital health interventions. He has been involved in the design and execution of serious games and gamified apps for health in European projects, as well as in gamified learning solutions for both healthcare professionals and patients in the pharmaceutical industry sector. His current research focuses on digital health interventions usage data analytics to improve engagement and behavioral change. His background includes a MSc in Computer Engineering at the University of Seville, Spain, and a MSc in Software Engineering at Cranfield University, United Kingdom. He also studied a short-term program in Healthcare Digital Marketing in Madrid, Spain, and several international gamification courses.

Dr. M. Househ is an Assistant Professor and a former Research Director at the College of Public Health and Health Informatics, King Saud bin Abdulaziz University for Health Sciences, National Guard Health Affairs, Riyadh, Kingdom of Saudi Arabia. He is also an adjunct Assistant Professor at the University of Victoria School of Information Science, Victoria, BC, Canada. As well, I am a consultant for the United Nations Development Program and the Editor-in-Chief of the *Journal of Health Informatics in Developing Countries.* In June 2015he was a Visiting Academic Researcher at the University of Oxford Nuffield, Department of Medicine, Centre for Tropical Medicine and Global Health. His primary research interests are around the use of information and communication technologies to empower patients and clinicians, specifically focusing on Social Media and Mobile Technologies in Healthcare in the promotion of public health practice and healthcare literacy.

Dr. E. Konstantinidis received the Diploma degree in electronic engineering from the Technological Educational Institute of Thessaloniki, in 2004, the MSc degree in medical informatics in 2008 from the Aristotle University of Thessaloniki, Greece, and the PhD degree in the Laboratory of Medical Physics of Medicine, School of Health Sciences, Aristotle University of Thessaloniki, Greece in 2015. His current research interests lie predominately in the area of Medical Informatics, particularly with respect to people with special needs and especially elderly. Recent research interests focus on intervention for elderly in the field of exergaming and IoT technologies. He has authored more than 30 publications in various international peer-reviewed journals and conferences. He is currently working in the H2020 project UNCAP.

Dr. S. Konstantinidis is an Assistant Professor in e-Learning and Health Informatics in the School of Health Sciences of The University of Nottingham in United Kingdom. He holds a PhD degree in Medical Sciences focusing on Medical Education Informatics from Aristotle University of Thessaloniki in Greece. He received his

Bachelor Degree in Computer Science from the University of Crete, Greece, in 2004 and his MSc in Medical Informatics from the Aristotle University of Thessaloniki, Greece, in 2007. He is a member at large of Global Health Workforce Council and he served as the Norwegian ambassador of OKFN. He has over 10 years of experience on EU, National, and interregional funded research projects in different roles including project coordinator's one (CAMEI—EU-FP7-CSA). He was a research associate at the Aristotle University of Thessaloniki (Greece) (2005–2012), and he was teaching at the Technological Educational Institute of West Macedonia (Greece). From 2012 until 2015, he was a researcher at NORUT—Northern Research Institute (Norway). He is active in publishing and contributing into the organization of academic events on Health Informatics and Medical/Health Education. His research interests include among other social media, content sharing, retrieval and repurposing, educational standards, collaborative e-learning, medical education informatics, virtual patients, linked open data, semantic web, serious games, human—computer interaction for seniors, gamification, exergames, and web-based health records.

Dr. L. Laranjo is a Medical Doctor who graduated from the Faculty of Medicine, University of Lisbon in 2007 and completed training in General Practice in 2014. She received a Junior Clinical Research Award from the Harvard Medical School, Portugal program in 2011 and finished the Master of Public Health from the Harvard School of Public Health in 2013. In 2015 she was awarded a PhD degree in Medicine from the Faculty of Medicine, University of Lisbon, for her thesis "Person-centered care and health information technology in Portugal—implications for chronic care and health quality improvement." She now works as a Postdoctoral Research Fellow at the Australian Institute of Health Innovation, Centre for Health Informatics, Macquarie University.

Dr. A.Y.S. Lau (BE, PhD) leads the most active consumer informatics research program in Australia, investigating new ways to improve health outcomes and patient engagement through the use of digital health. She has a national and international profile for her expertise in consumer informatics. Her long-term expertise lies in using informatics to design interventions for patients and consumers. This is an area which she strongly believes that will help us to achieve patient-centered care because of its ability to simultaneously empower individuals and to transform health service delivery. In the past decade, she has contributed a series of independent and original contributions to advance this area, focusing on the design and impact of consumer informatics, and the foundational understanding that underpins the way digital health is designed and utilized by consumers.

A. Leis is a Psychologist and holds an MSc degree in Health Psychology and Quality of Life of the Open University of Catalonia. For many years, she was the Assistant Director of the Department of Web Mèdica Acreditada at the Medical Association of Barcelona, participating in several European projects related to the use of health information by patients and health professionals on the Internet, the quality of health information, and the use of different filtering tools based on Semantic Web and text mining. Currently, she is a researcher in the Department of Experimental and Health Sciences at the Universitat Pompeu Fabra (UPF) in Barcelona, and she is developing her PhD research on the analysis of psychological profiles and mental disorders on Social Media.

Dr. M.A. Mayer (MD, MPH, PhD) is a Specialist in Family and Community Medicine. For many years, he worked for the Spanish National Health Service in the Internal Medicine Department of a general hospital and in Primary Care settings. He earned his PhD degree in Biomedical Informatics from the Universitat Pompeu Fabra (UPF) where he is an Assistant Professor in the Department of Experimental and Health Sciences. He was in charge of the Web Médica Acreditada Department of the Medical Association of Barcelona, an international web quality agency. He has participated in eight European projects since 2002, two of which are in progress, in programs on the Safer Internet Action Plan, DG SANCO, FP7, H2020, and Innovative Medicines Initiative. His research is focused on reusing EHR for research purposes, big data and health information management, and Social Media analysis.

Dr. M. Merolli is a researcher/academic in Participatory Health Informatics at the Melbourne Medical School, Health and Biomedical Informatics Centre (HaBIC), the University of Melbourne, Australia. His PhD developed a framework for research and practice to generate evidence about social media use in chronic disease management. He has a clinical health professional background with a Bachelors degree in Physiotherapy (honors). This background was a pivotal reason why he pursued participatory health informatics, interested in how patients and the health consuming public participate in their own health management using enabling technologies. He is an active member of the IMIA Participatory Health and Social Media Working Group, contributing to collaborative publications and panel discussions with colleagues. Outside of research, he regularly consults to health professionals and professional bodies on how digital media can improve clinical practice and operations and has developed social media training resources in this space.

Dr. O. Rivera is a Professor and researcher at the University of Seville, Spain. He got his computer science degree at the University of Seville. He worked as a technical staff in an ICT company and as a programmer at the University of Seville. His first research focuses were assistive technology and rehabilitation technology for people with disability. He has been a Lecturer/Professor at the University of Seville since 2005. He got his PhD in Industrial Computer Science in 2010. In 2012 he was a post-doc visiting research fellow at Norut IT, Norway. He is a member of a national working group to develop solutions and interventions to counteract the late effects of treatments of pediatric oncology since 2013. Currently, he is the principal investigator of an ehealth project. Also, he is participating on an H2020-funded smoking cessation project. His current research focus is behavioral change techniques and gamification applied to eHealth solutions.

Dr. S. Syed-Abdul is a citizen of India and a medical doctor. He is an Assistant Professor at Graduate Institute of Biomedical Informatics at the Taipei Medical University, Taiwan. He holds his Doctorate degree from National Yang Ming University in Biomedical Informatics; he earned his Master's degree in Telemedicine and eHealth from the University of Tromso, Norway, and graduated as medical doctor from St. Petersburg Medical Academy, St. Petersburg, Russia. He was project manager of various Health IT and mHealth projects like PWAS, Disease-Map, TrEHRT, LabPush, Sana–Mongolia, and Sana-Swaziland. His research interests are Big Data in healthcare, Analytics and visualization of Big Data, Social media in healthcare, mHealth, IoT in Health. He wants to focus on the MOOCs and social media to improve digital health literacy.

FOREWORD

The introduction of social media into healthcare seems insane to some people: good medicine depends on rigorous science, but social media are profoundly uncontrolled. Is it even possible that health and care could get anything valuable from this unmanaged "mob scene" of messages flying around the world??

The answer is yes, and it is the reason for this book. We have now seen that genuine clinical value can be brought to the table through social media, so for medicine to achieve its potential in the 21st century, we must understand both the possibilities and the risks. With that as foundation, we can design meaningful interventions.

Today, patients are recognized as a viable and valuable voice. The Institute of Medicine said 4 years ago, in *Best Care at Lower Cost*, that of the four pillars of a learning health system, #2 is "Patient/clinician partnerships" with "Engaged, empowered patients." Their wording evokes the term "e-patient" introduced in the 1990s by "Doc Tom" Ferguson (1946–2006), the visionary who saw that the Internet made new things possible for the flow of valuable information. It does not make patients oncologists but it lets us connect with information and with other patients.

Social media also serve the needs of clinicians who are overwhelmed with the thousands of papers published everyday, by letting them connect with other providers (as @KevinMD does) and letting them broadcast their message to patients anywhere (as pediatrician @SeattleMamaDoc and orthopedist @HJLuks do).

In the past, physicians and scientists were trained to only trust major "arteries" of information, clearly identifiable and highly predictable journals. Today, social media act like capillaries: countless pathways with no centralized control. Patients and clinicians of the future—and the best ones today—must understand how the pipelines of social media can bring value, and how to make the most of it. Read and learn.

Dave deBronkart

BIOGRAPHY

Dave deBronkart or cancer survivor "e-Patient Dave" deBronkart has become the world's best-known spokesman for the "e-patient" movement: empowered, engaged, equipped, enabled, and is especially articulate about how the Internet and social media are changing what is possible in healthcare. He has been online since CompuServe in 1989, where he co-led discussion forums with up to 40,000 people. That experience proved useful in 2007 when he was diagnosed with advanced kidney cancer; in addition to getting great medical care at Boston's Beth Israel Deaconess, he accepted the advice of his physician Dr. Danny Sands to join another online community: one composed of patients with his disease. With 32,000 Twitter followers, 2600 Facebook friends, and a Klout score of 80, today Dave uses social media, as well as his TED Talk and keynote speeches, to spread his message. Medicine respects him: the National Library of Medicine is capturing his blog in the History of Medicine; he was the Mayo Clinic's 2015 Visiting Professor in Internal Medicine, and he is in the Healthcare Internet Hall of Fame.

Already in 2008, the IMIA (International Medical Informatics Association) realized the need of strengthening research in the impact of Web 2.0 (aka social media) in medical informatics. Dr. Peter Murray facilitated the creation of the Web 2.0 Task Force at IMIA. This initiative rapidly consolidated and the IMIA Social Media Working Group was created, the activities of our group are based on the collaborative effort among many of us. The main reason behind the success of our working group (which it has won several awards from IMIA) is that it responds to an increasing demand of more knowledge about the use of social media in the health domain. One of our key characteristics in our working group is our multidisciplinary and multisectorial nature; this is not surprising since the demand of more knowledge and evidence-based guidelines in this area is shared across academic disciplines, healthcare practitioners, and patients.

When we started in 2008, it was quite easy to know most of the people working in aspects related to health social media. It was also feasible to read several dozens of papers and get a fair overview of this emerging field of research. Nowadays the story is quite different, there are thousands of people working in project related to health social media and hundreds of papers are published yearly. In this context it is very hard for people interested in research of health social media to get an overview of the current challenges and trends. This book is carefully designed for them. It can help people designing health social media interventions or defining their PhD research agenda. We do address issues such as serious social games for patient empowerment, big data, digital epidemiology, etc. After reading this book, you will be better equipped to design health social media strategies and research studies, whether you are an empowered patient or a health organization. This book is not a guideline on how to setup your own Social Media account, this is a book that goes beyond the "demo-tutoria," it will provide you well-explained examples and methodologies to use social media to improve the health of people.

Luis Fernandez Luque, PhD

Doha, Qatar

lluque@qf.org.qa

eHealth Researcher at Social Computing Group

Qatar Computing Research Institute,

Hamad bin Khalifa University

Qatar Foundation

Chairman of the IMIA Social Media Working Group

ACKNOWLEDGMENTS

A book of this scope would not have been possible without the contribution and support of the experts from multiple disciplines.

We acknowledge all the 20 enthusiastic authors, who made this book possible. We are fortunate to have counted with the contribution of eminent authorities in the field of social media and health from around the world.

As the editors, we also would like to specially thank the valuable participation of the following people who have help us in enriching this book:

Prof. Yu-Chuan (Jack) Li, Dean, College of Medical Science and Technology, Taipei Medical University, for his insight and useful comments.

Shwetambara V. Kekade, Masters student, Graduate institute of Biomedical Informatics, Taipei Medical University, for her partial contribution in the Introduction chapter.

Prof. Manuel Armayones, Faculty of Psychology, Open University of Catalonia, for his review of contents and valuable feedback.

Per Erlend Hasvold, mHealth, World Health Organization, reviewed contents and provided invaluable advice.

Dr. Kathleen Gray, from Health and Biomedical Informatics Centre, University of Melbourne.

Dr. Talya Miron-shatz, from Center for Medical Decision Making, Ono Academic College.

Dr. Hamish McLean, from School of Humanities, Languages and Social Science, Griffith University.

We are also grateful for the efforts and competence of our editors at Elsevier, Rafael Teixeira, Ana Claudia Garcia, and their colleagues for their professional input and guidance.

CHAPTER *1*

An Introduction to Participatory Health Through Social Media

S. Syed-Abdul[1], E. Gabarron[2,3], A.Y.S. Lau[4], and M. Househ[5]

[1]Taipei Medical University, Taipei, Taiwan [2]University Hospital of North Norway, Tromsø, Norway [3]The Arctic University of Norway, Tromsø, Norway [4]Macquarie University, Sydney, NSW, Australia [5]King Saud Bin Abdulaziz University for Health Sciences, Riyadh, Saudi Arabia

The use of social media for personal and health use by patients and clinicians is on the rise. This is part of a growing realization that social media can provide a platform for patients to gather information, explore options, and share their experiences. Social Media provides online platforms for interactions around various health topics relating to patient education, health promotion, public relations, and crisis communication. Social media includes various technological approaches such as blogs, microblogging (e.g., Twitter), social networking (e.g., Facebook and PatientsLikeMe), video- and file-sharing sites (e.g., YouTube), e-games, and wikis [1]. An important aspect of social media for health communication is to provide valuable peer, social, and emotional support for the general public and patients. Patients can share their experiences through discussion forums, chat rooms and instant messaging, or online consultation with a qualified clinician [2]. For example, social media can aid health behavior change such as smoking cessation, and "PatientsLikeMe" enables patients to communicate with other patients and share information about health issues [2].

Physicians are also using social media to promote patient health and education. Physicians tweet, create blog posts, record videos, and participate in disease-specific discussion forums focusing patient education. Such forums provide an important opportunity for physicians to distribute evidence-based information to counter inaccurate materials posted on the Internet. In some social media forums, the public is provided with an opportunity to participate in these discussions [3].

Participatory Health Through Social Media. DOI: http://dx.doi.org/10.1016/B978-0-12-809269-9.00001-3

There are a variety of benefits and limitations in using social media in healthcare. A perceived benefit is the accessibility and improved access to health information to various population groups, regardless of age, education, race/ethnicity, and locality compared to traditional communication methods. While these changing patterns may lessen health disparities, traditional inequalities and overall Internet access remain. Furthermore, variation in social media engagement according to personality traits, age, and gender suggests the need for ongoing scrutiny regarding equality of access and effectiveness for different users. Social media can be used to provide a valuable and useful source of peer, social, and emotional support to individuals, including those with various conditions/illnesses [2].

The primary limitations for social media are quality concerns and the lack of reliability of the health information. The large volume of information available through social media and the possibility for inaccuracies posted by users presents challenges when validating health-related information [4]. Several studies highlighted concerns about privacy and confidentiality, data security, and the potential harms that emerge when personal data are indexed. Social media users are often unaware of the risks of disclosing personal information online and communicating harmful or incorrect advice. As information is readily available, there is the potential of information overload [2]. Other concerns of the social media use for health are related to ethics. Some ethical challenges are still not answered such as how to deal with the presence of children and youth on the social media accessing and sharing health information and how to keep the privacy and confidentiality in the communication between healthcare professionals and patients through these channels. As there are no official guidelines proposing how to address the ethical considerations of using social media for health, it is recommended that researchers, healthcare professionals, and other stakeholders should carefully weigh the potential harms and benefits for the individuals in every case [5].

Social media brings a new dimension to health care, offering a platform used by the public, patients, and health professionals to communicate about health issues with the possibility of potentially improving health outcomes. Although there are benefits of using social media for health communication, the information needs to be monitored for quality and reliability, and the confidentiality and privacy need to be maintained. With increasing use of social media, there will be further opportunities and challenges for health care [2].

1.1 ORGANIZATION OF THE BOOK

This book explains how social media methodologies and platforms are used in healthcare. This book is organized into seven main chapters and starts with (1) introduction, (2) patient empowerment, (3) social media use by hospitals, (4) social media and health communication crisis during epidemics, (5) Big Data generated through Social Media, (6) Social Media and health behavior change, and ends with (7) gamification and behavioral change: techniques for health social media.

1.2 PATIENT EMPOWERMENT THROUGH SOCIAL MEDIA

Chapter 2, Patient Empowerment Through Social Media, offers a historical development of the patient empowerment movement and the role social media has in supporting this phenomenon. A recurring theme of the book, first articulated here, is the need to support patients to develop the capacity to take responsibility of their health. Understanding this fundamental concept and the potential of social media in supporting this movement is important for this book as it represents a paradigm shift of how social media can affect the relationship between patients and clinicians. The chapter explores this concept by illustrating the use of social media in both acute and long-term conditions, using sports concussion and diabetes as examples. It concludes with take-home messages on how we can use social media platforms to empower patients, taking into account their needs and their healthcare context.

1.3 SOCIAL MEDIA USE BY HOSPITALS AND HEALTH AUTHORITIES

Chapter 3, Use of Social Media by Hospitals and Health Authorities, examines social media use by hospital and health authorities, where it is estimated that 95% of US healthcare organizations use social media as part of various community engagement activities [6]. Social media has allowed healthcare institutions to increase visibility and improve their overall image as well as engage with patients. As a result, 12.5% of surveyed American healthcare organizations have successfully attracted new patients through the use of social media [7].

Griffis et al. [6] reviewed the use of social media across hospitals in the United States on four social media platforms: Facebook, Yelp,

Twitter, and Foursquare. Richter et al. [8] studied specifically how hospitals use social media and its opportunities to improve hospital services. Most of the hospitals surveyed used Facebook as a tool to educate patients, acknowledge staff, and share information about hospital recognitions such as awards. The authors found that the use of social media can help improve quality improvement efforts.

Dorwal et al. [9] studied the use of the social media messenger services, WhatsApp, in a hospital laboratory management department to improve communication among staff. Specifically, the authors found that there was a significant improvement in communication in the form of sharing photos, information about accidents, alerts, duty rosters, academic activities, and information from superiors. Johnston et al. [10] studied the use of WhatsApp to improve communication among emergency surgical teams. The authors conclude that WhatsApp was a safe communication tool in an emergency setting. These two studies show the use of social media, such as WhatsApp, as a hospital communication tool, where the results show benefits of social media use for communication. In 2015, Hui et al. [11] published a study of an online campaign using Facebook to promote attitudes for seeking mental health help.

One of the main challenges for healthcare organizations is related to understanding the meaningful use of social media sites by patients. Merely visiting a Facebook page or viewing a YouTube video does not signify meaningful use. Higher levels of interaction are needed between the patient and the healthcare institution, and studies to evaluate such interactions should be conducted. Other challenges include dedicating resources to design, maintain, advertise, and update social media sites. Healthcare organizations have not made sufficient investments in this area, and some are beginning to abandon or neglect their social media sites, especially, those organizations that do not see a financial, improved patient outcomes or reduced costs. Yet, the potential for community engagement through social media remains an opportunity for healthcare organizations to engage their communities and market their services.

In summary, there is evidence that social media use by hospitals and health authorities can lead to improvement in healthcare delivery. Although US hospitals have been successful in adopting social media in their healthcare service delivery model, other countries, especially those countries with low internet penetration and social media use, may not benefit from social media as an activity in promoting health-care. Furthermore, there are issues that relate to privacy and security concerns of patients, especially in countries where there are no health-care privacy laws that protect patient identities and patients from being manipulated by unscrupulous healthcare institutions.

Moving forward, to keep social media platforms running for the long term by hospitals and health authorities, a sustainable business model needs to be developed and implemented. Hiring of staff, techni-cal expertise, and resources is costly especially in a time where funding of healthcare organizations is shrinking globally. Not being able to generate revenue with the use of hospital resources has led to the mis-use, neglect, and the closing down of social media sites by hospitals and health authorities. For example, the Mayo clinic has implemented a successful business model that is generating revenue and has actually led to the expansion of the social media services it provides [12]. Other social media sites remain stagnant without dedicated staff to update them and are generally neglected.

There is a flurry of activities by hospitals and health authorities when using social media and the opportunities for growth both within the hos-pital and outside the hospital are real. Within the hospital, reducing costs, improving workflow, and improving communication are added benefits to the use of social media. More studies and examples of success-ful hospital implementations are needed. Outside the hospital, social media has been used to engage patients, obtain feedback on services, provide educational materials, and provide patients access to healthcare services. Although the long-term impacts on patient care are yet to be seen, thus far, social media's presence is increasing within the healthcare sector. The opportunity is ripe for future growth and sustainability.

1.4 SOCIAL MEDIA AND HEALTH COMMUNICATION CRISIS DURING EPIDEMICS

Building upon the concepts in the previous chapters, Chapter 5, Social Media and Health Communication Crisis During Epidemics, looks at how we can use data sources from social media for digital epidemiol-ogy and health crisis management. Today, electronic media and discus-sion groups are increasingly recognized as valuable sources of public health alerts. Gathering information from the Web now represents one important part of digital epidemiology, which comprises the idea that the health of a population can be assessed through digital traces. Crisis communications and emergency management have become more participatory, where social media represents an opportunity to broaden warnings to large population groups to support two-way communication. Through social media channels, information on disease activity or natural disasters can be distributed quickly to large groups of individuals. The chapter outlines an overview on approaches to digital epidemiology and health crisis management, illustrated with an example using Twitter and emerging technologies for the detection of outbreaks. The chapter concludes with ethical and legal aspects that need to be considered when using digital sources for crisis communication, population surveillance, and emergency management.

1.5 BIG DATA FOR HEALTH THROUGH SOCIAL MEDIA

There is an astonishing amount of social media data shared on the internet every minute [13]. Different social media platforms produce large amounts of data. For example, Facebook with its 1.44 billion active users shares approximately 31.25 million messages every minute [14]. It is estimated that by 2020 the amount of social media data will reach 44 zettabytes or 44 trillion GBs [13]. With the amount of social media data increasing at a rapid pace, the utilization of social media data to help decision makers in all industries, especially that of healthcare, remains unexplored.

The term Big Data has been used to describe the massive amount of data being shared and captured online, through sensor networks and information systems [15]. There has been much discussion on the use of Big Data and how it will transform the landscape of healthcare [16]. IBM Watson is a technology that utilizes social media data sources, as well as other data sources, to analyze healthcare data in the diagnosis and treatment of patients (IBM). In hospitals, Watson [17] can search electronic media, medical images, clinical research, the internet, and social media sources and provide a diagnosis and treatment plan with a 70−80% accuracy rate.

Much of the research on Big Data still remains at the conceptual phase with a scarcity of studies being conducted on the impact of Big Data on healthcare delivery and especially, patient care. In particular, there is scant research on the impacts Big Data is having on patient care with the use of social media data sources. Furthermore, much of the focus of Big Data and its use of social media has centered on

Public Health Surveillance. Making patient-related diagnosis with social media data is unreliable without the use of other related information sources coming from sources such as the electronic medical record.

Paul and Dredze [18] studied the use of social media, and specifically, Twitter for Public Health surveillance. The authors report that social media data can be used for Syndromic Surveillance to understand Geographic Behavioral Risk Factors and Geographic Syndromic Surveillance. Ram et al. [19] studied the prediction of asthma-related emergency department visits using big data. Yang et al. [20] conducted a pilot on the use of social media data as an early warning system for adverse drug reactions. The authors develop a system for pharmacovigilance leveraging a Latent Dirichlet Allocation modeling module to predict adverse drug reactions using social media data.

There are several limitations to the use of big data for social media. First, it is difficult to follow the progress of patients over time. Paul and Dredze [18] noted in their research that most of users relating to a specific disease only tweeted once about the public health issue, which makes it difficult to follow patients over time and predict their behavior. Second, using multiple social media platforms to collect data may cause data redundancy. Last, privacy and confidentiality of patients will be a concern with the use of Big Data, especially, as the Big Data technologies become more intelligent and are able to link a variety of different data sources that allow it to identify individuals.

In summary, Big Data use of social media generated information is rapidly increasing as more people share information online, and especially, when it takes a more automatic process with the increased use of the Internet of Things. In the near future, Big Data use of social media data alone will only be able to examine public health-related issues. It will be difficult, without linking a variety of different data sources, to longitudinally observe patients for diagnosis and treatment.

1.6 SOCIAL MEDIA AND HEALTH BEHAVIOR CHANGE

Chapter 6, Social Media and Health Behavior Change, discusses the importance of health behavior change for primary prevention (i.e., promoting and maintaining physical activity and a balance diet for preventing obesity, diabetes, or cardiovascular diseases); for chronic diseases management (i.e., improving disease knowledge and self-management behaviors); and also for treatment of mental health problems such as the management and treatment of anxiety and depression.

Numerous theories, models, and frameworks that attempt to explain the processes behind health behavior change are presented such as the Behavior Change Wheel, the Health Belief Model, the Social Cognitive Theory, the Transtheoretical Model, the Goal-Setting Theory, the I-Change Model, or the Social Change Theory. As example, this last model, the Social Change Theory, highlights the influence of group or social goals, norms and values, and it explains why health improvements are best achieved by altering norms at group or social level, rather than at the individual level.

This section exposed the tremendous potential, both at the population and individual levels, of the social media in the health domain; and it also suggested possible avenues for the use of social media in behavior change interventions; for disseminating interventions and recruiting participants due to its enormous reach; or to improve engagement. Specific challenges in the implementation of social media interventions for behavior change are mentioned too, such as the potential to link social networks with patient health records.

1.7 GAMIFICATION AND BEHAVIORAL CHANGE: TECHNIQUES FOR HEALTH SOCIAL MEDIA

Chapter 7, Gamification and Behavioral Change: Techniques for Health Social Media, addresses the use of games and gamification techniques for health and health social media. It gives an overview on how some scientific publications have focused on the disadvantages of the games for health (i.e., aggressive behavior, sedentarism, game addiction, etc.), while some other researchers have been able to use the engaging and intrinsically motivating game techniques for the benefit of health. The only intention of using all these gamification techniques is to engaging people in an environment where they learn and play simultaneously, leading to increase in their knowledge about health, and helping them to manage their health conditions.

Although the use of gamification techniques might not be effective in all health areas and is also unclear how players' personality and other psychological factors might affect their potential to engage in games for health and change behavior.

REFERENCES

[1] Househ M, Borycki E, Kushniruk A. Empowering patients through social media: the benefits and challenges. Health Informatics J 2014;20(1):50–8.

[2] Moorhead SA, et al. A new dimension of health care: systematic review of the uses, benefits, and limitations of social media for health communication. J Med Internet Res 2013;15(4):e85.

[3] Ventola CL. Social media and health care professionals: benefits, risks, and best practices. Pharm. Ther. 2014;39(7):491–520.

[4] Syed-Abdul S, Fernandez-Luque L, Jian WS, Li YC, Crain S, Hsu MH, et al. Misleading health-related information promoted through video-based social media: anorexia on YouTube. J Med Internet Res 2013;15(2):e30.

[5] Denecke K, et al. Ethical issues of social media usage in healthcare. Yearb Med Inform 2015;10(1):137–47.

[6] Griffis HM, et al. Use of social media across US hospitals: descriptive analysis of adoption and utilization. J Med Internet Res 2014;16(11):e264.

[7] Househ M. The use of social media in healthcare: organizational, clinical, and patient perspectives. Stud Health Technol Inform 2013;183:244–8.

[8] Richter JP, Muhlestein DB, Wilks CE. Social media: how hospitals use it, and opportunities for future use. J Healthcare Manag 2014;59(6):447–60.

[9] Dorwal P, et al. Role of WhatsApp messenger in the laboratory management system: a boon to communication. J Med Syst 2016;40(1):14.

[10] Johnston MJ, et al. Smartphones let surgeons know WhatsApp: an analysis of communication in emergency surgical teams. Am J Surg 2015;209(1):45–51.

[11] Hui A, Wong PW, Fu KW. Evaluation of an online campaign for promoting help-seeking attitudes for depression using a Facebook advertisement: an online randomized controlled experiment. JMIR Ment Health 2015;2(1):e5.

[12] Mayoclinic. About MCSMN. Available from: <http://socialmedia.mayoclinic.org/about-mcsmn/> [10.12.15].

[13] Kapco M. 7 staggering social media use by-the-minute stats. Social Media, Social Enterprise; 2015.

[14] Facebook. Available from: <https://www.facebook.com/help/> [March 2014].

[15] Basel Kayyali DK, Van Kuiken S. The big-data revolution in US health care: accelerating value and innovation. Available from: <http://www.mckinsey.com/industries/healthcare-systems-and-services/our-insights/the-big-data-revolution-in-us-health-care> ; 2013.

[16] Martin-Sanchez F, Verspoor K. Big data in medicine is driving big changes. Yearb Med Inform 2014;9(1):14–20.

[17] IBM Watson. IBM Watson Health. Available from: <http://www.ibm.com/smarterplanet/us/en/ibmwatson/health/> [12.12.15].

[18] Paul MJ, Dredze M. You are what you tweet: analyzing twitter for public health; 2011.

[19] Ram S, et al. Predicting asthma-related emergency department visits using big data. IEEE J Biomed Health Inform 2015;19(4):1216–23.

[20] Yang M, Kiang M, Shang W. Filtering big data from social media—building an early warning system for adverse drug reactions. J Biomed Inform 2015;54:230–40.

CHAPTER 2

Patient Empowerment Through Social Media

C.S. Bond[1], M. Merolli[2], and O.H. Ahmed[1,3,4]

[1]Bournemouth University, Bournemouth, United Kingdom [2]The University of Melbourne, Melbourne, VIC, Australia [3]Poole Hospital NHS Foundation Trust, Poole, Dorset, United Kingdom [4]The Football Association, London, United Kingdom

This chapter explores how social media can support patient empowerment and self-management. Although these concepts are generally linked with long-term (also called chronic) conditions, the value of social media in supporting more acute problems is also being explored. This chapter starts with a discussion of patient empowerment and then explores the use of social media to promote and support empowerment in both acute and long-term conditions (LTCs) using sports concussion and diabetes as examples.

2.1 EMPOWERMENT AND SELF-MANAGEMENT

Patient empowerment first started appearing in the academic literature in the 1980s as a model of healthcare that moved away from expecting patients to comply with the instructions set by their healthcare professionals, even when these required major and challenging lifestyle alterations. The empowerment model supported patients to develop the capacity to take responsibility for their own health and to have the knowledge necessary to be able to make decisions, the resources to carry out those decisions, and the experience to be able to evaluate their effectiveness [1].

Empowerment is most usually a concept linked to managing LTCs. LTCs are conditions that cannot be cured but can be controlled by the use of medication and/or other treatments and therapies [2] and are a global problem. It has been estimated that by 2020 they will account for 65% of all deaths [3].

In the United Kingdom, the National Health Service (NHS) has adopted a strategy to help empower patients living with an LTC who wish to develop better self-management capacity through commissioning an "Expert Patient Programme." This was based on the work done

Participatory Health Through Social Media. DOI: http://dx.doi.org/10.1016/B978-0-12-809269-9.00002-5

in the 1970s by Kate Lorig, who developed the Stanford University Chronic Disease Self-Management Program for people living with arthritis [4]. Rather than using professional health educators, this program introduced the concept of lay educators who have experience of the condition as leaders.

The NHS [5] describes an expert patient as being someone who:

- feels confident and in control of their life;
- aims to manage their condition and its treatment in partnership with healthcare professionals;
- communicates effectively with professionals and is willing to share responsibility for treatment;
- is realistic about how their condition affects them and their family;
- uses their skills and knowledge to lead a full life.

The program, which is paid for by NHS and is offered free of charge to participants, is based around 6, 2.5-hour sessions run by two tutors who also have LTCs, and covers general health topics including:

- dealing with pain and extreme tiredness;
- coping with feelings of depression;
- relaxation techniques and exercises;
- healthy eating;
- communicating with family, friends, and healthcare professionals;
- planning for the future.

This approach has not been without its critics, including Wilson [6] who questions the power issues inherent in the system. The argument has been made that there needs to be a strategy to challenge healthcare professional assumptions towards people with chronic conditions, as well as one to change the behaviors of people living with chronic conditions. The efficiency of expert patient programs has also been questioned [7], with a lack of involvement beyond the patient, to family and other social support being questioned.

Whilst the criticisms may be justified to some extent, the aim of empowerment described by Anderson and Funnell [8] promotes a shift from a model where healthcare professionals use education as a tool to increase patient compliance to a model where it is used to increase the patient's freedom, ability to think critically, and act autonomously. It has been suggested that positive health outcomes tend to result from multidisciplinary support systems and better overall organization of care, which sits in contrast to more traditional reactive models of healthcare of dealing with illness if or when it arises [9].

Closely aligned to the concept of the expert patient is "shared decision making," a model developed to explore effective ways of involving patients in making treatment decisions in complex scenarios (e.g., deciding the best approach to treatment for newly diagnosed breast cancer [10]). In the United Kingdom, shared decision making is a clearly stated health policy through the government's "No decision about me without me" initiative [11]. Whilst this is seen as fundamental to achieve safe and effective healthcare provision, it has been found to not yet be fully embedded in clinical practice [12].

Building social support networks is a feature of the expert patient approach that has been suggested leads to improved self-management support and can positively influencing lifestyle change [13]. The Internet, especially social media, has enabled new networks to be formed independently. These networks may include family, peers with the same condition or health interest, and friends online.

An alternative model to the expert patient is participatory medicine, a concept initially proposed by Dr. Tom Ferguson [14] who described two approaches to healthcare: Industrial Age Medicine and Information Age Medicine. This has been updated by Bond [15], with this model supporting an alternative approach to patient empowerment (Fig. 2.1). Rather than a model that is centered on healthcare professionals, it

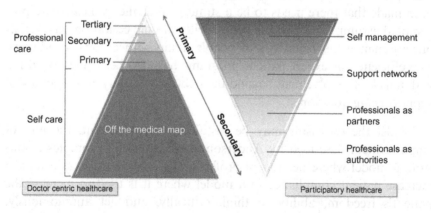

Figure 2.1 Participatory healthcare model.

proposes a model where self-management and peer-to-peer support are key features.

Some patient groups may feel more comfortable, or find it more achievable, to create their own health networks online. Health status, isolation, and stigmatizing conditions are some of the common reasons for people with LTCs developing an online presence. Social media has provided new and diverse opportunities to connect in this way [16]. Research in this area is continually growing; in one commonly cited report, Fox [17] suggested that if given the opportunity, people living with LTCs like to take advantage of social media in managing their health.

Using social media in health self-management represents a paradigm shift and can be seen as a progression from searching for static online health information provided on traditional web pages. The potential of social media tools (e.g., social networks, blogs, video sharing platforms, etc.) to turn online health self-management into the user-centric, participatory, and engaged endeavor identified by Eysenbach in the early days of social media has been realized [18].

Today, responsive healthcare providers have also started to see the value of engaging and collaborating with connected patients through various social media platforms to foster self-management support to achieve positive health outcomes [19,20]. This lies at the heart of shared decision making between patients and healthcare providers and leads to autonomy, improved communication, greater self-efficacy, and better patient satisfaction [21].

2.1.1 Examples: Acute and Long-Term Conditions
2.1.1.1 Acute—Social Media and Sports Concussion
Social media has opened new opportunities for the empowerment model and for participatory healthcare. Whilst much of the focus has been on living with LTCs, there have also been innovations in how patients can be empowered to meet acute care needs such as those arising from sports injuries. This is a fast-moving and dynamic area, and the use of social media has significant potential to enhance the well-being of patients with a multitude of short-term ailments. One of the areas where its use has been explored is sports concussion (mild traumatic brain injury), a condition which has received significant recent attention in both the mainstream media [22] and the scientific literature [23].

Sports concussion is generally a short-lived condition. The majority of cases resolve in 7–10 days [24], although in some cases symptoms can persist [25] and there can be serious consequences from cumulative injuries [26]. Research into trends of adoption, use, and disengagement amongst online health communities [27] has found that with regards to disengagement, transitioning in and out of online communities was noted and was deemed to be acceptable and not an indication of the failure of the community. Patients may only require a transient, short-lived engagement with online communities or social media formats in order to meet any unmet health needed they have related to their condition. Although the duration of this support in concussion is not as long lasting as that for LTCs, it can still have a significant impact and play a positive role in the recovery and return to function for an individual.

Countries with public health systems are frequently overburdened [28] and patients may not always have immediate access to assist them with managing a short-term condition such as sports concussion. For countries with private healthcare systems, patients may not have the necessary healthcare insurance or feel that they want to pay privately for their short-term condition. Empowering patients through online peer support [29] and online consultations [30] may constitute a viable method of helping them to recover more safely from concussion than they would do independently.

2.1.1.1.1 Examples in the Field of Sports Concussion

The role that social media plays in relation to sports concussion has been investigated in recent years with several descriptive studies outlining its use. The preliminary work in the field examined concussion-related Facebook groups [29]. Inclusion and exclusion criteria were used to identify 17 eligible Facebook groups with the postings on these groups evaluated using content analysis methodology. The majority of individuals were using Facebook to communicate with others who had also sustained a concussion, with relatively few examples seen of healthcare providers interacting with patients online. A subsequent study exploring the use of Twitter [31] saw 1000 tweets collated over a 7-day period and then categorized using a coding scheme. The majority of the tweets (33.2%) were news stories related to concussion, however, a sizeable proportion of tweets (26.8%) were sharing personal information related to an experience with concussion. Only a small number of tweets (7%) were seen to downplay the severity of concussion.

Concussion-related YouTube content was evaluated by Williams et al. [32] with the authors selecting 100 concussion-related videos with the highest view counts. Content analysis was again used to categorize each video with the most common categories being depiction of a sporting injury (37%) and news reports (25%). Whilst many of the videos came from news agencies (51%) or laypersons (30%), only 4% of videos originated from government agencies (2%), medical centers (1%), or professional associations (1%). Image-sharing social media platforms (Pinterest, Instagram, and Flickr) were subsequently assessed by Ahmed et al. [33], with a cross-platform coding scheme applied to these three websites. In keeping with the previous studies in the area, content analysis was used for the images retrieved from each site. Images were also cross-referenced against the best practice concussion management guidelines [24], to assess whether content was reflecting gold-standard care postinjury. Although it was not possible to determine this for many images, for those where it was possible the overwhelming majority (91%) were adhering to the best practice guidelines.

The common themes emerging from across these platforms are a clear pattern of patients seeking out others who have experienced the same condition as them, and seeking support and interacting with them. It is also noticeable that healthcare providers and healthcare organizations are not leading this information exchange relating to concussion, and that it is a patient-led process. There were relatively few examples of misinformation being shared by patients in these studies; however, given the difficulties the medical community has had in transferring concussion-related information to the general public [34] then it is surprising that social media has not been more readily adopted by major medical organizations for this purpose. A cohesive, cross-platform approach with concussion information tailored and targeted to the needs of the local population could certainly assist patient empowerment and fully utilize the capacity of social media.

2.1.1.1.2 Considerations for Patients with the Platforms Used
Each of the major social media platforms listed in Table 2.1 provides various levels of functionality and enables different aspects of user involvement, and this has obvious implications for the way that patients can engage with them to obtain support. Although they may

Table 2.1 Studies Comparing Social Media Usage Related to Sports Concussion			
Social Media Platform	Date of Study	Main Findings	Potential Implications
Facebook [29]	2009	Most individuals used Facebook to share a concussion-related story. Smaller proportions were seeking advice or offering concussion advice	Peer-to-peer interaction using Facebook could serve as a useful knowledge transfer mechanism
Twitter [31]	2010	Twitter was well used for disseminating concussion information. News-related stories were the most commonly shared forms of information	The rapid information exchange on Twitter means it could provide an excellent means for organizations to disseminate best practice messages
YouTube [32]	2012	The majority videos were of a sporting injury, predominantly from news stories but also a sizeable proportion of user-generated incidents	Healthcare and educational organizations should utilize this potentially powerful medium to share concussion-related advice
Image-sharing websites [33]	2013	Primary purpose of images shared was to depict a concussion-related scene. There were very few instances where inappropriate postconcussion management was seen	Image-sharing websites could be used as an adjunct to traditional information sharing approaches
Smartphone apps [35]	2013	There was a large diversity in the quality and the range of concussion smartphone apps on the market, and a consumer checklist was generated to help choose an appropriate smartphone app	Smartphone apps constitute a valuable means of assisting the initial assessment of sports concussion, but only if the app is of a sufficient quality

have been restricted in their scope when they first emerged, all of these social media platforms now enable static text, videos, and pictures to be shared. It is important that the strengths (and also the limitations) of the platforms are understood in order to maximize their potential to assist patients. For example, although Twitter may allow users to share information rapidly in a brief tweet, longer discourse and interactions will be limited by the constraints of the platform.

From a commercial aspect, most social media platforms now have a degree of sponsorship and advertisements attached to them. Twitter, Facebook, YouTube, and Instagram all have sponsored links present, and it is important that there is a clear distinction between clinical and commercial information if patients are to gain the most from using social media to assist their recovery. Ethical issues relating to the use of

Facebook for healthcare support have been highlighted [36], and these need to be safely overcome if sports medicine is to comprehensively embrace the use of social media as a regular and effective adjunct to medical care in this manner.

Sports medicine at the elite level is a lucrative market with financial pressures associated with it [37], and at a community/mainstream level there is scope for misleading patients online who are looking for a "quick fix" to their sporting injury. This is especially pertinent with regards to smartphone apps in sports medicine, which have been shown to vary in their quality and functionality [38,39]. Organizations in the field of sports medicine need to be aware of the responsibility of their clinicians to the general public and monitor these new methods of clinician—patient interactions accordingly.

2.1.1.1.3 Scope for Future Use of Social Media in Sports Medicine (and Other Short-Term Conditions)

This discussion provides a brief example of the existing and potential uses of social media for sports concussion, but there is also much scope for patients to be empowered in other areas of sports medicine (and for other short-term medical conditions). A key starting point in this exploration is stakeholder engagement to ascertain the needs and demands of the target demographic and to identify potential barriers to usage. This process also needs to include the clinicians treating these individuals [40] to maximize the efficacy of social media use in this manner.

An aging population in many developed countries (e.g., the United Kingdom [41]) and a focus on promoting physical activity for all ages [42] mean that in the future the successful management of sports injuries will become more important, and innovative methods to help better management of these injuries could become an important public health issue. In comparison to more established fields of medicine, the branch of sports medicine is relatively new and therefore is still establishing itself, and there exists an opportunity for the sports medicine profession to capitalize on the benefits that social media can offer patients. The high-profile nature of professional sport and the large numbers involved in recreation and social sport means that there is likely to be a commercial interest in such developments too, meaning regulation is needed to prevent those seeking to obtain a capital advantage from inadequate/clinically unproven innovations in this field.

2.1.2 Long-Term Conditions—Self-Management, Empowerment, and Peer Support

Active online communities are in existence providing support and patient empowerment for multiple conditions (e.g., cancer [43] and diabetes [44]). One of the challenges of self-management in this regard is when the patient's right to make their own decisions about their health differs from the opinion of their healthcare professionals [45]. Traditionally patients' expertise has been seen as being based on their experience of living with their condition rather than their knowledge of their condition [46]. This view is starting to be challenged [47] and it is acknowledged that people living with LTCs spend the majority of their time managing their conditions. For example, people living with diabetes are suggested to spend an average of 3 hours per year interacting with healthcare professionals [48].

The majority of research into self-management has explored how interventions can be developed that will empower people and support effective self-management. Not as much is known about what people living with an LTC want to help support their self-management strategies. A study undertaken in 2012 [49] studied online support forums for people living with diabetes to explore the relationship people had (and wanted) with their healthcare professionals. They specifically explored what people:

- considered their role in condition management to be
- considered their doctor's role in managing their condition to be
- saw as positive elements of their interactions with medical staff
- found problematic in their interactions with medical staff.

This study used a qualitative approach to analyze posts made on public discussion boards. This raised ethical research issues around the public/private nature of online discussion boards. A study into the ethical issues in using data from online discussion boards [50] found that participants were aware that their contributions to these discussions are available to the public. There were, however, mixed feelings about if, and how, this information should be used for research. Whilst most participants in the study were happy about their information being used for research, some individuals expressed a desire to maintain control over the future use of this information. Differences of opinions were seen with regards to the use of direct quotes, with some participants wanting their contributions acknowledged if they were used as quotes, whilst others wanted to remain anonymous. As a result of this, recommendations have been made that online contributions should be treated as a unique type of research data, named "personal health text" by the authors, and anonymized before being presented in any public form [50]. This is the approach that the study [49] being discussed took.

Most of the contributors to the forums considered that they were responsible for their own care. The role of healthcare professionals was ambiguous, however, with some contributors seeing their value mainly as gatekeepers to the services they needed (such as prescriptions and blood tests). Others valued doctors who were willing to work in partnership, both learning with and learning from them. Contributors were not necessarily unhappy when they and their doctors disagreed, as long as this disagreement was approached from a position of mutual trust.

Healthcare professionals were an important source of information for the people in this study, however, they were often not the main source of information nor was information from healthcare professionals seen as privileged or more reliable than other sources. Contributors to online forums often signpost their peers to information sources they think will be useful, as well as offering their own advice. People who were confident in their self-management capabilities liked to select information that they felt helped them to manage their condition from multiple sources, including their peers, rather than rely on any one source.

One aspect of self-management found in these online forums was the need to be able to manage their interactions with healthcare professionals. The need for a good relationship was widely recognized and patients were advised to be clear, confident, and assertive. Patients wanted to be recognized as being equal partners in the relationship and to be able to talk to professionals on a level standing. If patients did not feel their relationship was good they were happy to support others to find ways of coping and achieving their aims, including allowing the doctors to think their advice was being followed but actually "doing their own thing."

2.1.3 Therapeutic Affordances of Social Media in Long-Term Condition Management

Research has assisted our understanding of health outcomes reported from social media use by providing evidence of their therapeutic impact on self-management of people with LTCs. However, to advance understanding in this area even more, diving deeper into "how" social media use might precipitate health effects is helpful [16,51]. What is it about interacting with social media that might underpin the empowerment outcomes such as validation, confidence, and self-esteem for different people? What does social media therapeutically afford people who use them?

"Affordances" are a complex concept originating from psychology. The concept has often been used to study human−computer interaction and essentially looks to explain the actions that are possible when an individual interacts with objects in their environment [52]. It also takes into account how possible interactions are conveyed to the person based on their past experiences, needs, wants, and goals [53]. When exploring health outcomes resulting from social media use, the concept may be useful to help understand how people may feel empowered from using social media. The platform, its interface, and the service it can provide, along with the person self-managing their condition, their goals, attitudes, preferences, motivations, and experiences all need to be considered [54]. Using this model helps to piece together that different people self-managing various LTCs will approach, engage with, and understand various social media platforms differently.

Merolli et al. [51] put forward five therapeutic affordances of social media in managing LTCs. Here their relationships to patient empowerment are explored. These were based on findings from a survey of people living with chronic pain who use social media as part of self-management.

1. Self-presentation:
   How an individual controls, how they present themselves online, and how much people are willing to disclose and share about their identity and/or condition.
   People living with chronic pain who use social media have reported that they felt a sense of control and autonomy over their interactions through social media. Suggestions as to how this assists them include being able to control social interactions and interacting with others on one's own terms. The empowering effect of being able to dictate control over one's identity disclosure has been echoed in earlier research [55].

2. Connection:

   How people connect with others through social media, exchange information, support each other, and mitigate isolation.

   Support available through social media has been reported to be pivotal to feeling understood and validated. A sense of connection to others who share similar experiences through social media is said to be emotionally cathartic and motivating. People have suggested that interactions shared with others and an ability to share information lead to greater sensations of self-worth. As previously discussed, much of the power of connection centers upon support. Studies have explained the strength of personal connections through social media as well as the sense of community that can develop as a result [56].

3. Exploration:

   Social media offering the ability to search for information online that may help with disease management.

   In 2008, Eysenbach introduced the concept of apomediation through social media [18], meaning that social media "guides" people towards useful information to manage their health. In the context of patient empowerment, it has been reported that through the guiding/filtering effect of social media, useful pain management resources could be sourced. These in turn often leading to other useful resources. The result has been suggestive of providing patients with greater confidence to open discussions with healthcare professionals. Furthermore, "learning" from information found using social media has also been observed. Some have commented that the ability to explore information has added to their knowledge base considerably—with the adage "knowledge is power" ringing true here.

4. Narration:

   People living with LTCs sharing experiences with illness and the emotionally cathartic role this can play.

   The therapeutic effect of the narrative has been well supported in social media use for LTC management research. The narration therapeutic affordance has been reported to statistically correlate most strongly to reports of improved health outcomes (e.g., "improved relationships with others" and "participation in social activities") [57]. This was particularly associated with use of social network site, blog, and discussion forum. Similarly to "connection" of individuals, people living with LTCs have suggested that being able to pass on disease-specific management information to others is an uplifting experience and tied to a sense of personal satisfaction and worth. On the other hand, learning from the experiences of others is also positively linked to empowerment, in particular to understanding and validation. Mo and Coulson [55] elaborated on this suggesting that learning from others' experiences helps people makes sense of their own condition and improves their sense of control over it.

5. Adaptation:
   The way social media can give users the ability to alter or adapt self-management behaviors based on disease-specific needs at different points in time.

   Merolli (2015) was able to demonstrate statistical correlations between adaptation (especially surrounding more frequent use of social media) and health outcomes. Other studies have also reported this effect [55,58]. Empowerment may be recognized through the autonomy social media enables, allowing individuals to change their usage patterns dependent on whether their condition is stable, flared, or otherwise. Also, social media can provide a means to stay connected to friends and family when conditions require hospitalization or social isolation [51].

2.2 TAKE-HOME MESSAGES

It is reasonable to assume that in the coming years, new social media platforms will emerge with new functionality to assist (and also entertain) society [59]. Simultaneously, existing social media platforms will also evolve and adapt in order to sustain the interest and engagement of their user base. As such, clinicians and healthcare innovators need to be aware of how patients are interacting with these new platforms and which ones show promise and potential to be used for patient empowerment and for peer-to-peer support self-management. When identifying the use of new social media platforms for healthcare, the balance between the functionality of the platform, the needs of the population, and the demands of the condition must be considered (Fig. 2.2). Throughout this design process, clinicians need to work collaboratively with their peers from other forms of healthcare, including their patients, to develop ways in which they can utilize these new platforms, and harness them to assist the recovery from short-term conditions the management of long-term conditions, and to enhance the quality of life of people living with long-term conditions.

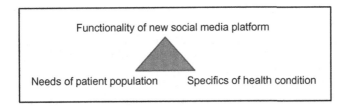

Figure 2.2 Considerations when adopting new social media platforms for healthcare.

Key Points:

• Participatory healthcare is moving beyond early models of empowerment to a model where people living with LTCs are seen as experts and shared decision makers—in both their experience of living with their condition and their knowledge of their condition.

• Effective participatory self-management has moved beyond seeking information to increase patients' levels of adherence to healthcare professionals and patients working together to agree how a condition is best managed.

• Social media has the power to support people with short-term health needs as well as long-term (chronic) conditions

• Responsive healthcare providers see the value of engaging and collaborating with connected patients through various social media platforms to foster self-management support to achieve positive health outcomes.

REFERENCES

[1] Funnell MM, Anderson RM, Arnold MS, Barr PA, Donnelly M, Johnson PD, et al. Empowerment: an idea whose time has come in diabetes education. Diabetes Educ 1991;17 (1):37–41.

[2] Department of Health. Long term conditions compendium of information, third ed. Leeds; 2012.

[3] Lorig K, Ritter PL, Plant K, Laurent DD, Kelly P, Rowe S. The South Australia Health Chronic Disease Self-Management Internet Trial. Health Educ Behav. 2013;40(1):67–77.

[4] Donaldson L. Expert patients' and self-management. London: Department of Health; 2006.

[5] Department of Health. The Expert Patients Programme (EPP) NHS Choices: Department of Health. Available from: <http://www.nhs.uk/NHSEngland/AboutNHSservices/doctors/Pages/expert-patients-programme.aspx>; 2015.

[6] Wilson PM. A policy analysis of the Expert Patient in the United Kingdom: self-care as an expression of pastoral power. Health Social Care Commun 2001;9(3):134–42.

[7] Greenhalgh T. Chronic illness: beyond the expert patient. BMJ 2009;338(7695):629–31.

[8] Anderson RM, Funnell MM. Patient empowerment: myths and misconceptions. Patient Educ Couns 2010;79(3):277—82.

[9] Fenlon D, Frankland J, Foster CL, Brooks C, Coleman P, Payne S, et al. Living into old age with the consequences of breast cancer. Eur J Oncol Nurs 2012;.

[10] Charles C, Gafni A, Whelan T. Shared decision-making in the medical encounter: what does it mean? (or it takes at least two to tango). Soc Sci Med 1997;44(5):681—92.

[11] Department of Health. Liberating the NHS: No decision about me, without me. London; 2012.

[12] Joseph-Williams N, Elwyn G, Edwards A. Knowledge is not power for patients: a systematic review and thematic synthesis of patient-reported barriers and facilitators to shared decision making. Patient Educ Couns 2014;94(3):291—309.

[13] Funnell MM. Peer-based behavioural strategies to improve chronic disease self-management and clinical outcomes: evidence, logistics, evaluation considerations and needs for future research. Fam Pract 2010;27:17—22.

[14] Ferguson T. Understand e-patients 2002. Available from: <http://www.doctom.com/slide-shows/>.

[15] Bond CS. Telehealth as a tool for independent self-management by people living with long term conditions. In: Maeder AJ, Mars M, Scott RE, editors. Global Telehealth 2014. Durban: IOS Press; 2014.

[16] Merolli M, Gray K, Martin-Sanchez F. Health outcomes and related effects of using social media in chronic disease management: a literature review and analysis of affordances. J Biomed Inform 2013;46(6):957—69.

[17] Fox S. The social life of health information. Washington, DC: Pew Research Center Publications; 2011. Available from: <http://pewresearch.org/pubs/1989/health-care-online-social-network-users>.

[18] Eysenbach G. Medicine 2.0: social networking, collaboration, participation, apomediation, and openness. J Med Internet Res 2008;10(3).

[19] Seeman N. Web 2.0 and chronic illness: new horizons, new opportunities. Healthcare Q 2008;11(1):104—8 10, 4.

[20] Martin-Sanchez F, Lopez-Campos G, Gray K. Chapter 11—Biomedical informatics methods for personalized medicine and participatory health. In: Sarkar IN, editor. Methods in biomedical informatics. Oxford: Academic Press; 2014. p. 347—94.

[21] deBronkart D. From patient centred to people powered: autonomy on the rise. BMJ 2015;350(148).

[22] Times TNY. In Europe, echoes of America as concussions spur debate. Available from: <http://www.nytimes.com/2014/04/06/sports/in-europe-echoes-of-america-as-concussions-spur-debate.html?_r=0>; 2014.

[23] Mrazik M, Dennison C, Brooks B, Yeates K, Babul S, Naidu D. A qualitative review of sports concussion education: prime time for evidence-based knowledge translation. Br J Sports Med 2015;49(24):1548—53.

[24] McCrory P, Meeuwisse W, Aubry M, Cantu B, Dvorak J, Echemendia R, et al. Consensus statement on concussion in sport: the 4th International Conference on Concussion in Sport held in Zurich, November 2012. Br J Sports Med 2013;47(5):250—8.

[25] Leddy J, Sandhu H, Sodhi V, Baker J, Willer B. Rehabilitation of concussion and post-concussion syndrome. Sports Health Multidisc Appr 2012;4(2):147—54.

[26] Patricios J, Kemp S. Chronic traumatic encephalopathy: Rugby's call for clarity, data and leadership in the concussion debate. Br J Sports Med 2014;48(2):76—9.

[27] Massimi M, Bender J, Witteman H, Ahmed O. Life transitions and online health communities: reflecting on adoption, use, and disengagement. In Proceedings of the 17th ACM conference on computer supported cooperative work & social computing; Baltimore, MD: ACM; 2014. p. 1491–501.

[28] BBC News. NHS pressure worsens as key targets missed. Available from: <http://www.bbc.co.uk/news/health-34790100>; 2015.

[29] Ahmed O, Sullivan S, Schneiders A, McCrory P. iSupport: do social networking sites have a role to play in concussion awareness? Disabil Rehabil 2010;32(22):1877–83.

[30] Vargas B, Channer D, Dodick D, Demaerschalk B. Teleconcussion: an innovative approach to screening, diagnosis, and management of mild traumatic brain injury. Telemed e-Health 2012;18(10):803–6.

[31] Sullivan S, Schneiders A, Cheang C, Kitto E, Lee H, Redhead J, et al. What's happening? A content analysis of concussion-related traffic on Twitter. Br J Sports Med 2012;46(4):258–63.

[32] Williams D, Sullivan S, Schneiders A, Ahmed O, Lee H, Balasundaram A, et al. Big hits on the small screen: an evaluation of concussion-related videos on YouTube. Br J Sports Med 2014;48(2):107–11.

[33] Ahmed O, Lee H, Struik L. A picture tells a thousand words: a content analysis of concussion-related images online. Phys Ther Sport 2016;. Available from: http://dx.doi.org/10.1016/j.ptsp.2016.03.001.

[34] Weber M, Edwards M. Sport concussion knowledge in the UK general public. Arch Clin Neuropsychol 2012;27(3):355–61.

[35] Lee H, Sullivan SJ, Schneiders AG, Ahmed OH, Balasundaram AP, Williams D, et al. Smartphone and tablet apps for concussion road warriors (team clinicians): a systematic review for practical users. Br J Sports Med 2015;49(8):499–505.

[36] Ahmed O, Sullivan S, Schneiders A, Anderson L, Paton C, McCrory P. Ethical considerations in using Facebook for health care support: a case study using concussion management. PM&R 2013;5(4):328–34.

[37] Anderson L, Jackson S. Competing loyalties in sports medicine: threats to medical professionalism in elite, commercial sport. Int Rev Soc Sport 2013;48(2):238–56.

[38] Ahmed O. The smartphone app "Rotator Cuff Injury/Strain" by Medical iRehab. Br J Sports Med 2015. Available from: http://dx.doi.org/10.1136/bjsports-2015-095036.

[39] Wong S, Robertson G, Connor K, Brady R, Wood A. Smartphone apps for orthopaedic sports medicine —a smart move? BMC Sports Sci Med Rehabil 2015;7:23.

[40] Ahmed O, Sullivan S, Schneiders A, Moon S, McCrory P. Exploring the opinions and perspectives of general practitioners towards the use of social networking sites for concussion management. J Primary Health Care 2013;5(1):36–42.

[41] UK Parliament. The ageing population: key issues for the 2010 Parliament. Available from: <http://www.parliament.uk/business/publications/research/key-issues-for-the-new-parliament/value-for-money-in-public-services/the-ageing-population/>; 2010.

[42] NHS Choices. The importance of exercise as you get older. Available from: <http://www.nhs.uk/Livewell/fitness/Pages/activities-for-the-elderly.aspx>; 2015.

[43] Bender JL, Jimenez-Marroquin MC, Jadad AR. Seeking support on Facebook: a content analysis of breast cancer groups. J Med Internet Res 2011;13(1):e16.

[44] Greene JA. Online social networking by patients with diabetes: a qualitative. J Gen Int Med 2011;26(3):287–92.

[45] Goodyear-Smith F, Buetow S. Power issues in the doctor–patient relationship. Health Care Anal 2001;9(4):449–62.

[46] Badcott D. The expert patient: valid recognition or false hope? Med Health Care Philos 2005;8(2):173−8.

[47] Bond CS, Hewitt-Taylor J. How people with diabetes integrate self-monitoring of blood glucose into their self-management strategies. J Innov Health Inf 2014;21(2).

[48] Department of Health. Supporting people with long term conditions: an NHS and social care model to support local innovation and integration. London; 2005.

[49] Hewitt-Taylor J, Bond CS. What E-patients want from the doctor−patient relationship: content analysis of posts on discussion boards. J Med Internet Res 2012;14(6) e155

[50] Bond CS, Ahmed OH, Hind M, Thomas B, Hewitt-Taylor J. The conceptual and practical ethical dilemmas of using health discussion board posts as research data. J Med Internet Res 2013;15(6):e112.

[51] Merolli M, Gray K, Martin-Sanchez F. Therapeutic affordances of social media: emergent themes from a global online survey of people with chronic pain. J Med Internet Res 2014;16 (12):e284.

[52] Zhao Y, Liu J, Tang J, Zhu Q. Conceptualizing perceived affordances in social media interaction design. Aslib Proc New Inform Perspect 2013;65(3):289−303.

[53] Norman D. The design of everyday things. Basic Books; 2002.

[54] Bradner E. Social affordances of computer-mediated communication technology: understanding adoption. In CHI'01 extended abstracts on human factors in computing systems, Seattle, WA. 634111: ACM; 2001. p. 67−8.

[55] Mo PK, Coulson NS. Online support group use and psychological health for individuals living with HIV/AIDS. Patient Educ Counsel 2013;93(3):426−32.

[56] Huh J, Liu LS, Neogi T, Inkpen K, Pratt W. Health Vlogs as social support for chronic illness management. ACM Trans Computer−Human Interact 2014;21(4):1−31.

[57] Merolli M, Gray K, Martin-Sanchez F, Lopez-Campos G. Patient-reported outcomes and therapeutic affordances of social media: findings from a global online survey of people with chronic pain. J Med Internet Res 2015;17(1):e20.

[58] Stellefson M, Chaney B, Barry AE, Chavarria E, Tennant B, Walsh-Childers K, et al. Web 2.0 chronic disease self-management for older adults: a systematic review. J Med Internet Res 2013;15(2):e35.

[59] Anderson J, Raine L. Digital life in 2025. Pew Research Centre; 2014. Available from: <http://www.pewinternet.org/2014/03/11/digital-life-in-2025/>.

CHAPTER *3*

Use of Social Media by Hospitals and Health Authorities

S.A. Adams
Tilburg University, Tilburg, The Netherlands

3.1 INTRODUCTION

Health practitioners, educators, and authorities increasingly recognize the added value of including social media such as (micro)blogs, social networking sites, and media sharing sites in various processes/types of health information exchange. Different types of social media can be used for different purposes and the scale of reach to potential audiences, speed with which information is shared and responded to, as well as the ease of combining different information formats (i.e., text, image, video), are considered to be advantageous for both distributing and collecting information about health, care, and healthy practices [1]. These applications are increasingly considered to be important tools for improving patient engagement and the delivery of patient-centered care [2]. They are expected to help increase patient participation not only in the personal care process but also in processes related to the governance of care, such as patient representation in the political arena, activism, or quality improvement.

At the same time, while the promises of social media are widely recognized, use in the health systems of many countries is sporadic and hesitant, at best. Professionals and institutions (including health authorities) are concerned about, for example, possible ethical and legal issues related to social media use, both with one another and for purposes of patient engagement [3]. Concerns range from keeping professional distance when posting *to* social media to providing an appropriate response to signals about health risks that may be gleaned *from* all the information being openly exchanged [4]. While there are some codes of conduct already available, more guidance for professionals and institutions is necessary in order to facilitate trust in these technologies.

Participatory Health Through Social Media. DOI: http://dx.doi.org/10.1016/B978-0-12-809269-9.00003-7

In this chapter, I examine how hospitals and public health authorities (i.e., health ministries/government departments, public health organizations, health educators, and inspectorates) are currently using social media and outline three basic uses of social media for patient and public engagement: distributing information, collecting information, and exchanging information. I highlight the *perceived* advantages and disadvantages associated with each type of use and emphasize the word perceived here, because actual experience may be different in practice. This is especially the case with many perceived disadvantages, which appear to be worries about the unknown that can often be addressed in practice with practical examples and suggestions coming from the field.

As I have argued elsewhere [1,3], because there are so many types of social media that can be used in different ways in different context, teasing out the issues is more nuanced if we reflect on individual applications (rather than "social media" as a whole). Where possible I therefore give concrete examples of which type of social media application is used for a particular purpose and which perceived advantage or disadvantage is linked to which application type.

3.2 THREE USES OF SOCIAL MEDIA

Generally speaking, there are three basic uses of social media for patient and public engagement: (1) Distributing information *to* patients and the public, (2) Collecting information *from* patients and the public, and (3) Facilitating interaction, both *among* patients and *between* patients and providers [4]. Whereas the first two basic uses are somewhat distanced from the generally accepted properties of "social" media (i.e., mass public information *exchange*) in that they reflect primarily unidirectional information flows, the third basic use is closer to the activities most commonly associated with the various social media platforms discussed throughout this book. It should be noted that the increasing integration of different types of social media onto one platform or enabling syndication (RSS) with the push of a button makes these three practices somewhat indistinguishable in practice. That is, merely sending information may lead to responses that enable collecting or exchanging information. Nonetheless, for reasons of clarity, I distinguish the three uses here.

3.2.1 Distributing Information to Patients and the Public

Policy changes over the last 20–30 years have increasingly redefined patients into consumers, not only of health services but also of health-related information [5,6]. This is part and parcel to what has been deemed a New Public Management approach to health policy [7], whereby a results-driven culture for public services (including health-care) is promoted and citizens are redefined as "customers" of those services. Patients and potential patients are provided with various types of information about health, care, treatment, and provider performance so that they can use this information to make relevant choices when deciding about their care. Both institutions and professionals are expected to adjust their services in order to meet customer needs [8]. Patients-as-consumers are considered, in this respect, to be levers in the governance of health providers, which arguably makes government and professionals more accountable to the public, increases transparency and efficiency of performance, improves quality of care, decreases expenses, and gives citizens more control in their own care process.

3.2.1.1 Health Institution Profiling

This means that there is an increasing pressure on health institutions to profile or market themselves as the best choice for patient care [3]. Care institutions take to the web to show their innovative practices, patient-centered approach, specialized expertise, high-quality care, etc. In many cases the marketing or media departments of these institutions function as gatekeepers, sending controlled messages through, for example, tweets, much as they would a press release for newspaper or television. They may also establish a platform (such as a weblog) where they can post news of the day, promote activities, and respond to news items or comment on current events. These new platforms are seen as the new "business card" of the organization, whereby the main advantage is the potential reach of what is posted there and the speed with which information can be distributed along these channels.

Organizations are concerned, however, about how quickly they may lose control over their online image—either by unknowing errors made by employees or by possible public reaction or redistribution of posted information. In the Netherlands, academic medical centers have implemented strict guidelines for their employees with regard to posting information on blogging sites or via Twitter (often using examples from the commercial sector—e.g., Coca-Cola and Nike—as "good practice" and then adjusting these to make them more healthcare specific). Such policies remind professionals that when they post information on the web in their professional context, this also reflects on the organization and encourage them to think about the associations that can be made with a given user name, as well as to keep professional posts and online personae separate from private posts. Some organizations initially prescreened posts by employees, but the rapid growth in popularity of, for example, Twitter has made this an unmanageable task.

Nonetheless, organizations do attempt to retain control. While many social media platforms contain interactive functions that allow users to post comments, mark "favorites," or "like" a particular page or group on many platforms (especially media sharing sites such as YouTube and Flickr), these can be cut off or hidden in the interest of preempting and preventing possible undesired reactions, such as negative comments or votes. Professional organizations have stepped in to help their members learn to engage with social media and to retain control over the information exchange.

Although most professional groups have learned some rules quickly, for example, not to accept "friend" requests from patients on social networking sites and to be aware of who is (or might be) following what they write on a Twitter feed or personal weblog, they also realize that the boundaries between professional and personal are not always as clear-cut as guidelines might suggest. As is further discussed below, because they cannot anticipate all the possibilities, professionals are also concerned about possible unintended effects of their actions (or inaction) on their relationships with patients. For example, can not accepting a "friend" request on a networking platform such as Facebook upset the patient and therefore disrupt an established trust relationship? Not yet having answers to these types of specific concerns often inhibits their use in general.

3.2.1.2 Health Authority Messaging

Health authorities such as safety inspectorates or institutes of public health also use social media as part of their profiling strategy. However, this is not driven by market pressure to show a competitive advantage as in the case of health institutions; rather it is part of a political strategy to legitimate decisions made and work being done. This is especially true in countries with a national health program (where healthcare is a public good) such as the United Kingdom and in regulated-competition markets that combine financing for healthcare, such as in the Netherlands. More importantly, however, is the use of social media to keep the public informed about news and current events. This may be about policy decisions that affect the work of the authority in question (e.g., new laws that reappropriate funds or restructure division of responsibilities), but it may also be about specific events that affect the public health (e.g., natural disasters, outbreaks of disease). Especially institutes of public health find it important that the general public has a reliable source of information in such situations. They can therefore send short, controlled text messages that mix updates specific to the situation with more general reminders about how to act or respond to the situation. For example, in a European case where a train accident led to the release of toxic chemicals in the area, residents of the area were informed about the nature of the accident, followed by a reminder to close doors and windows and to stay inside until a release message was posted.

3.2.2 Collecting Information From Patients and the Public

A second use of social media by health institutions and authorities is to gather and collate health "sentiments" [9] as they become more prevalent among the general public. This may be content-driven, politically driven, or as a form of quality control. The most prominent form of content-driven data collection through social media is disease surveillance [10] but also includes crisis management (see also chapter: Patient Empowerment Through Social Media).

3.2.2.1 Disease Surveillance

In the past, monitoring trends in illness relied on the reporting activities of health professionals to public health organizations (such as the Centers for Disease Control in the United States). This meant that trends could only be detected if and when patients presented with symptoms. Overviews were not present in real-time, but often during or even after the height of a viral wave or epidemic.

With social media, mention of symptoms such as fatigue can be an initial indicator of illness, which can be confirmed later through increase in severity or new symptoms. If the data generated via social media are mined for these types of words and examined for the context in which they are used, then the identification of a trend and process of monitoring its development are moved forward in time, accelerating the process of near-real-time disease surveillance. While the role of data mining tweets and Google searches to identify and follow outbreaks of influenza is one the most prominent examples of public health authority use of social media for disease surveillance [11,12], there are also efforts to create programs for data mining other types of social media platforms.

One example is the Netherlands Institute for Public Health and the Environment, which is developing programs to mine posts to community platforms (forums) in real-time and create visualizations such as word clouds. This is an added dimension to disease surveillance in that it moves beyond monitoring outbreaks of acute disease and can enable public health educators to trace the sentiments of and other aspects associated with living with a longer-term condition such as cancer or self-managing a chronic disease.

The speed with which monitoring can take place and trends can be identified is the greatest perceived advantages of this use of social media for gathering information about public health. Public health authorities can take action more quickly, providing targeted information or intervening at specific locations, if necessary. Those personally afflicted with a virus or disease may benefit directly from early intervention and possibly curbing the spread of such an illness arguably leads to general health benefits for the entire population.

The primary perceived disadvantage is found in the technical capacity. Although technical capacity is growing, many government bodies, especially in European countries, where "home-grown" systems are more common, do not yet have the resources for optimal data mining. They often must build these from scratch and even when the technical know-how and financial resources are available, it still takes time to determine the approach and algorithms to be used and then write these into a program.

3.2.2.2 Quality Control/Incident Reporting

A different type of signal that can potentially be identified through following postings to social media platforms are those relating to the quality and safety of provided care. Social media applications have also enabled the establishment of, for example, rating and recommendation sites, where patients can review various aspects of their care, including institutional performance and interactions with professionals [13] and the (perceived) effectiveness and/or side effects of medications and treatments [14]. These websites provide a wealth of information on the quality of service delivery that hospitals and care providers can use to improve daily practice and authorities such as inspectorates (state survey agencies) can use to signal potentially risky situations. In preparing for inspection visits, they may incorporate this information into an institution's dossier and ask follow-up questions.

The perceived advantage of this use of social media is the ability to pick up signals about, for example, the quality of care, which might otherwise remain undisclosed. It is a low-threshold approach to gathering such information and enables more people to participate on a much greater scale. Making such information publicly available also arguably increases responsiveness to customer concerns—transparency of information about performance means that institutions and professionals must answer demands for improvements in quality of care. In national (i.e., publicly funded) health systems such as the United Kingdom, these sites are indeed often seen as an instrument of responsiveness to the tax payers. The website Patient Opinion [15] has booked notable success in improving the quality of care and the Care Quality Commission (CQC) has examined how it might use that information in its institutional review work, as well. Similarly, in countries such as the United States, these sites are becoming ever more common for pressuring care institutions to be transparent about performance.

However, in countries such as the Netherlands, there has been much critique of these sites, with professionals and authorities questioning whether the web is the most appropriate outlet for this information [16]. Within the Dutch system, structures such as complaint departments of hospitals, regional organizations with patient representatives, and patient associations that act as advocates tend to be seen as more appropriate methods for capturing patient comments and concerns, although this is also changing, with the introduction of such sites by patient associations themselves.

Where physicians are reviewed online by name, there is concern about "naming and shaming" and a negative bias in reviews. Studies in the Netherlands have shown this concern not to be founded [13,17], but it is the case that the "staying power" of information on the Internet and the tendency of physicians and institutions to provide an offline response to these types of reviews may lead to a different representation of events on the web that inadequately reflects what happened in practice.

More generally, *all* mining of patient experiences (whether in tweets about health sentiment or reviews of quality of care) has been questioned by critical sociologists, such as Ref. [18], who argue that there is an unprecedented commodification of patient experience and opinion. This implies that others benefit from having patients share experiences, with little reflection on the consequences for the individual.

3.2.3 Facilitating Interaction

Different social media applications can be used to facilitate interaction, especially within existing relationships, in a number of ways. This section examines three types of social media use for facilitating interaction: within specific professional–patient relationships, in doctor-to-patient (D2P), and patient-to-patient (P2P) communities, recruiting for research and for general patient/public education.

3.2.3.1 D2P and P2P Communities

Hospitals in the Netherlands are beginning to use blog-based community applications to engage with patients. The idea behind this approach is to help them to manage their own health information [19] and help them connect with their peers to exchange information [20]. These platforms enable different types of interaction, for example, personal health records, online doctor consultation, and peer discussion forums (now called communities), available through one log-in function. Depending on the platform in question, friends and family who provide care for a loved one may also be able to access parts or all of the information.

Increasingly these platforms also contain connectivity functions for issues related to care. For example, relatives of an elderly person in need of help keeping house may be able to connect through such a platform with someone looking for a volunteer opportunity. Especially for elderly patients, such platforms are expected to improve social support and integration in the community.

The primary perceived advantage of interactive interfaces where patients (and their loved ones) can access and manage personal health and medical information and interact with professionals and/or patient peers is that of patient empowerment. They may provide medical, emotional, and lifestyle support for, for example, self-management. This is expected, at least over the long term to reduce the burden on healthcare systems by enabling people to do more for themselves [21]. A second perceived advantage, especially in D2P communities is the combination of peer insights and medical knowledge in one place [20]. Institutions do view such platforms as more reliable when patients and physicians together contribute information to the site, with the idea that the professional provides correct, authoritative information that patients can communicate to one another in their own words.

Perceived disadvantages, however, are limits to patient skill in engaging with such sites. They may not understand all of the functionalities of the site or be limited by their illness or prevented by other factors from capitalizing on what is offered there. Moreover they may not always lead to the expected increase in feeling empowered [19]. The return on investment is therefore not yet as high as expected, neither in social advantages such as patient empowerment nor in economic advantages, such as decreased burden on care systems.

3.2.3.2 Recruiting for Research

Recently there has been more attention for how social networking sites such as Facebook can be used to recruit for research [22,23], highlighting the ease with which people can join in, as well as the secondary benefits of social connectivity and interaction for individual patients. Institutions are turning to platforms such as Facebook and LinkedIn, where they can set up groups focused on specific target groups. Hospitals may host interactive "community" platforms where patients can interact with professionals and other patients.

One perceived advantage of this approach is ease of access to potential research subjects, increased participation in studies, and decreased rates of attrition as a study progresses. To increase uptake and use of these platforms, health researchers may lower the threshold of use by allowing users to log-in using an existing social media account.

This incorporates a trade-off, however, as studies from both inside and outside of healthcare have discussed potential threats to user privacy (see, e.g., Refs. [24–27]) on social networking sites such as Facebook. High profile news of Facebook's own disputable practices [28] also disrupts trust in this platform as an option for use in relation to health/medical information exchange. This aspect of using these platforms is the largest perceived disadvantage in relation to the research purpose. Researchers who recruit via such sites rarely have full awareness of how these sites manage the data exchanged there. This means that they do not have control over how the data is further used by the site, for example, being sold to third-parties that may use it for commercial interests.

Physicians also find it difficult to assess how far they can (or should) go in answering their *own* patients' questions if the conversation is being held in the public sphere. While they realize that this could be a service to the patient and a good move toward patient-centered care, they are also concerned about how this might possibly breach confidentiality, which is a nonnegotiable aspect of patient consultation. Similarly, administrators are concerned about the degree to which both professionals and the institutions could be held legally liable for any information exchanged (or, indeed for failing to act upon a signal). These concerns could lead to discouragement at the institutional level of using social media, even in the interest of patient-centered care.

3.2.3.3 General Patient/Public Education

Media sharing sites make it increasingly easy to open out institutional practices to the general public. This means that information about hospital processes (such as admission) and certain types of (surgical) procedures or treatments can be more widely distributed among the general public or to patients prior to their being admitted for a given care condition. Whereas institutions have explored the option of using virtual reality interfaces and games for this purpose [29] (see also chapter: Social Media and Health Behavior Change), they are increasingly turning to Flickr, YouTube, and even Twitter.

Notable examples are the world's first live-tweeted heart operation by the Memorial Hermann Northwest Hospital in Houston, Texas in 2012 [30] and a similar event in the Netherlands [3] in 2013. The Dutch event began with short promotional videos with snippets of the physician explaining the problem and corrective procedure to the patient on YouTube, followed by physicians, nurses, and the patient tweeting about the pending operation in the days leading up to it. During the operation, a physician not involved with the procedure tweeted about the event and posted informative pictures to Flickr and afterward, the patient tweeted about his recovery. The patient also asked peers to share experiences about what to expect and in turn and fielded their questions (together with the surgeon) about various aspects of the procedure. In both cases, followers also created their own version, for example, by posting a narrative to the story creation site Storify and of course, these items were also picked up in more "traditional" media, such as television and newspapers.

Such examples show how different social media platforms can converge around a single event and may be used to exchange different types of information with different audiences. Where people pick up health information and "make it their own," this can be an effective tool for patient and public education. It also enables tailoring information to specific health conditions and situations.

Perceived advantages of initiating these types of activities are the ability to provide better education to the public about what happens in the hospital, as well as to show the current state of the art in medical techniques. This is of course good not only for the institution in terms of reputation management and profiling but also for patients who feel they are able to ask for and receive exactly the information they are looking for—answers to their specific questions formulated in a language they understand.

Perceived disadvantages are, as mentioned above, issues related to privacy and confidentiality. Especially in the Dutch case the identity of the patient was known weeks in advance, but many safeguards were put in place with respect to informed consent, but also how to deal with possible unexpected difficulties during the procedure. Moreover, this raises the question of where to establish boundaries when bringing a personal interaction with a patient into the public sphere. If other patients ask questions, what is the physician's duty to respond and to what degree is she/he held responsible/liable for the information exchanged?

There are also concerns about the aforementioned idea of commodification of patient experiences. Institutions and physicians gain notoriety via their exchanges on these platforms, which means that under the pretense of patient-centered care, there is much (economic) benefit of health institutions, public health authorities, and possibly even other actors.

3.3 DISCUSSION

This brief review shows that health authorities and institutions at different levels are beginning to explore possible uses of various social media platforms for gathering, disseminating, and exchanging medical and health-related information with patients and the general public. It also shows that while the advantages of using these platforms are widely recognized, many unknowns and perceived disadvantages currently inhibit use or lead to questions that professionals, managers, and government authorities have difficulty in addressing.

One of the most notable perceived disadvantages was the possible threat to individual privacy and related effects on the treatment relationship. Many professionals are aware of the "fluid" nature of privacy policies for different types of social media (especially social networking sites) and understand that nothing is ever 100% protected. Nonetheless, it remains important to raise awareness about organizational and professional inability to protect sensitive user data if such sites are used. Especially when institutions are asking users to provide or exchange health information through *any* of these platforms (including logging into a private site via a commercial social media account such as LinkedIn, Twitter, or Facebook), they should ensure that users are aware of the hidden aspects of data management and control. Institutions may build their own in-house systems without commercial log-in and otherwise should provide clear information of this aspect of data exchange as part of the informed consent process.

Another concern is about how to define boundaries, especially as exchanges between different actors become more public and distributed across different types of platforms. While the health benefits of such publicness [31] have been well-argued and demonstrated for specific groups of patients [32], engaging the public in general and specific patients or groups through these media still requires forethought. Professional guidelines attuned to the context—not only health culture of a particular country but also intended patient or public audience, specific social media platform, and reason for use—will help those initiating such information activities know where to draw the line. Not only professionals need such assistance, but also health authorities and institutions can use guidance. Moreover there is a need for these actors to help educate patients, especially providing full disclosure when they are monitoring information streams for health signals or actively soliciting information from patients about their personal health and care.

3.4 CONCLUSION

As with any new technology, use in practice requires a period of sorting things out and learning-by-doing, in which users must learn to weigh the (perceived) benefits and detriments, and develops strategies for addressing these concerns. Use in any context involves a number of trade-offs that must be considered with an eye on preventing (unintentional) errors and fostering appropriate use. In healthcare the stakes are higher and the trade-offs vary in relation to a number of contextual factors such as type or purpose of use, specific social media application used, and intended audience for the communication or exchange. Insufficient protection of patient and professional rights, inattention to ethical considerations and errors, however unintentional, could adversely impact not only the relationship between patients and professionals but also between citizens and those institutions responsible.

Generally there are benefits to engaging patients and the public through social media, which point to arguments favoring its use, also in healthcare settings, as well as arguments against taking a hard line to guard against potential detriments, such as only using social media for purposes of promoting an institution (which is currently the case for many health organizations). In many cases, not only professionals but also care institutions (i.e., at the managerial level) and health authorities need more guidance in learning the "ins and outs" of social media for patient and citizen engagement. This will help to secure more trust in social media use so that we can reap the benefits of posting, gathering, and exchanging information about various aspects of health and care in these everyday interactions.

REFERENCES

[1] Adams SA. Revisiting the internet reliability debate in the wake of 'web 2.0': an interdisciplinary literature and website review. Int J Med Inform 2010;79:391–400.

[2] Rozenblum R, Bates DW. Patient-centred healthcare, social media and the internet: the perfect storm? BMJ Qual Saf 2013;22:183–6.

[3] Adams SA, van Veghel D, Dekker L. Developing a research agenda on ethical issues related to using social media in healthcare: lessons from the first Dutch Twitter heart operation. Camb Q Healthc Ethics 2015;24:293–302.

[4] Adams SA, van de Bovenkamp H, Robben P. Including citizens in institutional reviews: expectations and experiences from the Dutch Healthcare Inspectorate. Health Expect 2013;18:1463–73.

[5] Clarke J, Newman J, Smith N, Vidler E, Westmarland L. Creating citizen-consumers. London: Sage; 2007.

[6] Hardey M. Doctor in the house: the internet as a source of lay health knowledge and the challenge to expertise. Sociol Health Ill 1999;21:820–35.

[7] Gray A, Harrison S, editors. Governing Medicine: Theory and Practice. Maidenhead: Open University Press; 2004.

[8] Cordella A. E-government: towards the e-bureaucratic form?. J Inform Technol 2007; 22:265–74.

[9] Desai T. Tweeting the meeting: an in-depth analysis of twitter activity at Kidney Week 2011. PLoS ONE 2012;7.

[10] Bernardo TM, Rajic A, Young I, Robiadek K, Pham MT, Funk JA. Scoping review on search queries and Social Media for Disease Surveillance: a chronology of innovation. J Med Internet Res 2013;15:e147.

[11] Terry T. Twittering healthcare: social media and medicine. Telemed eHealth 2009;15: 507–10.

[12] Paul MJ, Drezde M. A model for mining public health topics from Twitter. Health 2012;11:6–16.

[13] Adams SA. Sourcing the crowd for health services improvement: the reflexive patient and "share-your-experience" websites. Soc Sci Med 2011;72:1069–76.

[14] Adams SA, Using patient-reported experiences for pharmacovigilance? In: Beuscart-Zephir MC, Jaspers M, Kuziemsky C, Nøhr C, Aarts J, editors. Context Sensitive health informatics: human and sociotechnical approaches. Studies in health technology and informatics, vol. 194; 2013. p. 63–68.

[15] Patient Opinion website. Available from: <http://www.patientopinion.org.uk>; [last accessed 18.12.15].

[16] Adams SA. Post-panoptic surveillance through healthcare rating sites: who's watching whom? Inform Commun Soc 2013;16:215–35.

[17] Geesink R, Koolman X. Does ZorgKaartNederland.nl give a representative impression of the quality of care providers? [available only in Dutch]. Amsterdam: Talma Institute; 2013.

[18] Lupton D. The commodification of patient opinion: the digital patient experience economy in the age of big data. Sociol Health Ill 2014;36:856–69.

[19] Tuil WS, Verhaak Chris M, Braat DDM, de Vries Robbé PF, Kremer JAM. Empowering patients undergoing in vitro fertilization by providing Internet access to medical data. Fertil Steril 2007;88:361–8.

[20] Vennik F, Adams SA, Faber M, Putters K. Expert and experiential knowledge in the same place: patients' experiences with online communities connecting patients and health professionals. Patient Educ Couns 2014;95:265–70.

[21] Harris R, Wathen N, Wyatt S. Configuring health consumers: health work and the imperative of personal responsibility. New York, NY: Palgrave Macmillan; 2010.

[22] Ramo DE, Prochaska JJ. Broad reach and targeted recruitment using facebook for an online survey of young adult substance use. J Med Internet Res 2012;14:e28.

[23] Chu JL, Snider CE. Use of a social networking web site for recruiting canadian youth for medical research. J Adolesc Health 2013;52:792–4.

[24] Parry M, Harvard researchers accused of breaching students' privacy. The Chronicle of Higher Education. Available from: <http://chronicle.com/article/Harvards-Privacy-Meltdown/128166/>; July 10 2011 [last accessed 15.12.14].

[25] Zimmer M. "But the data is already public": on the ethics of research in Facebook. Ethics Inf Technol 2010;12:313–25.

[26] Wyatt S, Harris A, Adams SA, Kelly SE. Illness online: self-reported data and questions of trust in medical and social research. Theor Cult Soc 2013;30:128–47.

[27] Fuchs C. Web 2.0, prosumption, and surveillance. Surveill Soc 2011;8:289–309.

[28] Goel V, Facebook tinkers with users' emotions in news feed experiment, stirring outcry. The New York Times, June 30 2014, B1.

[29] Adams SA. Use of "serious health games" in health care: a review. Stud Health Technol Inform 2010;157:160−6.

[30] Laird S. World's first live-tweeted open-heart surgery is a success. Available online: <http://www.mashable.com/2012/02/23/tweeted-open-heart-surgery/>; 2012 [last accessed 15.12.14].

[31] Jarvis J. Public parts. New York, NY: Simon and Schuster; 2011.

[32] Frost JH, Massagli MP. Social uses of personal health information within PatientsLikeMe, an online patient community: what can happen when patients have access to one another's data. J Med Internet Res 2008;10:e15.

CHAPTER 4

Social Media and Health Crisis Communication During Epidemics

K. Denecke[1] and S. Atique[2]
[1]Bern University of Applied Sciences, Bern, Switzerland [2]Taipei Medical University, Taipei, Taiwan

By 2020, there will be 6.1 billion smartphone users, so it is time to get serious about digital epidemiology. Researchers have already started to develop methods and strategies for using digital epidemiology to support infectious disease monitoring and surveillance or understand attitudes and concerns about infectious diseases. But much more needs to be done to integrate digital epidemiology with existing practices and to address ethical concerns about privacy.

4.1 DIGITAL EPIDEMIOLOGY AND DISEASE SURVEILLANCE

A variety of factors such as population movements, behavioral changes, or food production are responsible for the continuous emergence of infectious hazards. Diseases such as *severe acute respiratory syndrome* (SARS), avian influenza, or bioterrorism caused by the deliberate release of biological agents, all represent new challenges for outbreak alert and response worldwide. Only early detection of disease activity, followed by a rapid response, can reduce the impact of epidemics and prevent harm caused by disease outbreaks [1]. The World Health Organisation defines a disease outbreak as the "occurrence of cases of disease in excess of what would normally be expected in a defined community, geographical area or season." An outbreak may occur in a restricted geographical area or may extend over several countries. It may last for a few days or weeks, or for several years [2]. A single case of a communicable disease long absent from a population, or caused by an agent (e.g., bacterium or virus) not previously recognized in that community or area, as well as the emergence of a previously unknown disease, may also constitute an outbreak and should be reported and investigated immediately after its occurrence.

Participatory Health Through Social Media. DOI: http://dx.doi.org/10.1016/B978-0-12-809269-9.00004-9

Surveillance systems support the management and early detection of disease activity [1]. Traditional surveillance systems that rely on reported diagnoses from laboratories, doctors, or hospitals are well established in all EU countries. While traditional systems can recognize trends over a long time period, and ensure a public health response to identified risks, new emerging threats such as SARS, human cases of avian influenza, might remained unrecognized. Furthermore, despite the development of new approaches for the detection of previously unknown threats (e.g., monitoring of syndromes, death rates, drug prescriptions), these are still insufficient, because signals leading to a public health alert can originate from other sources.

Today, electronic media and discussion groups are increasingly recognized as valuable sources of public health alerts. Awareness of diseases achieved through first-hand observations and "word of mouth" can influence people's behavior and reduce the risk of an outbreak and the number of infected people [3]. Therefore, gathering information from the Web now represents one important part of *Epidemic Intelligence* [1]. Epidemic intelligence combines all efforts for systematic health event detection by providing a conceptual framework into which countries may complete their public health surveillance system. The objective of Epidemic Intelligence is to complement traditional surveillance systems by going beyond traditional public health surveillance and incorporating new official and unofficial sources of structured and unstructured information [1].

The more general concept of *digital epidemiology* [4] comprises the idea that the health of a population can be assessed through digital traces, in real time. Consider the following example: many people suffer from flu every year and many of them search for relevant information in the internet, and share their health problems with others online. In this way, a description of their symptoms, time-stamped and even geo-tagged, is available through search logs, social networks, or other social media tools. Therefore, the internet provides a rather detailed picture of the health of the population, coming from digital sources, through all of our connected devices, including smartphones.

Once an outbreak has occurred, it is crucial for health experts and volunteers to have efficient means for health risk and crisis communication and assessment. Crisis communication is an ongoing process associated with the exchanging information of opinions on a crisis and the coordination of resources including equipment, personnel, and information to avoid or reduce harm and for coordinating resources during a crisis [5,6]. It also includes the strategy to make people's behavior more rational that they could make informed decisions. The Organisation for Economic Cooperation and Development (OECD) is claiming in a report that "Social media are revolutionizing communication" [5]. They report three ways to use social media in crisis management: (1) as situation awareness tool, (2) as state communication tool, or (3) as a platform for dynamic interaction. Natural disasters such as the 2010 Haiti earthquake or flood in Thailand revealed already the utility of internet-based social media for risk and crisis communication [7]. In these contexts, social media represented an opportunity to broaden warnings to large population groups. OECD acknowledges a "great potential to support two-way crisis communication at low cost and with high efficacy" [5].

In the following sections, we will describe social media data sources and their content with respect to a use in digital epidemiology and health crisis management. We further provide an overview on approaches to digital epidemiology and analyze their strengths and weaknesses in a SWOT (strength, weakness, opportunities, and threats) analysis. Concrete experiences from one project will be summarized. Given the progress in technology, it is often easy to implement new tools for digital epidemiology. However, ethical and legal aspects must be considered carefully. Some of these aspects will be presented at the end of this chapter.

4.2 SOCIAL MEDIA DATA AND EPIDEMICS

What can we find in social media and in the internet with respect to disease activity? Who is reporting on what and in which manner? How can social media support crisis communication and risk assessment? This section tries to answer these questions. As described before, digital epidemiology relies upon sensors, which are humans, that are recognizing and reporting disease activity on social media or leave other digital traces in the web.

Social media are internet-based applications that enable people to share their own information via the internet. This form of communication is more common than ever before and has gained unprecedented popularity around the world through social networking websites like Facebook or microblogging websites like Twitter. The trend is also recognizable in the healthcare field, where people are accessing websites for medical advice, joining patient communities, and are posting information about their own health status [8].

Social media data includes various kinds of publicly available content that is produced by end-users, rather than by the operator of a website. Medical social media data is a subset of the social media data space, in which the interests of the participants are specifically devoted to medicine and health issues [8]. More specifically, with medical social media data we refer to web-based narrative text and data that contain medical content which was written by individuals (potential patients), physicians, or other healthcare professionals.

4.2.1 Data Sources
In general, the content in medical social media is characterized by a mixture of expert knowledge, layman knowledge or experiences and empirical findings. We can distinguish different social media tools where this content is distributed.

Social networking sites with health-related content enable people with similar interests to connect. More specifically, patients who suffer from diseases can share health data in order to empathize with each other or learn about treatments, physical exercises, or medications other patients are consuming in order to improve their health status. For example, PatientsLikeMe (https://www.patientslikeme.com/, last accessed 17.11.2015) is a social network for patients that allows them to share health-related experiences and compare treatments. The community currently comprises more than 350,000 patient members (November, 2015). Over 2500 conditions are reported in the platform. Data is collected in a rather structured manner: for the various features such as quality of life or single symptoms, categories are predefined (e.g., quality of physical life on a scale of 4 between best and worst, see Fig. 4.1). Access to health social networks is often restricted to members, i.e., only registered members can connect to others and read through their content. For applications in digital epidemiology, posted messages would need to be collected and analyzed automatically. Given the restricted access, this is difficult if not impossible, also due to legal issues (see Section 4.6). However, social networks offer the opportunity to be used in crisis communication, either for coordinating emergency services and volunteers, or to share information inside a community.

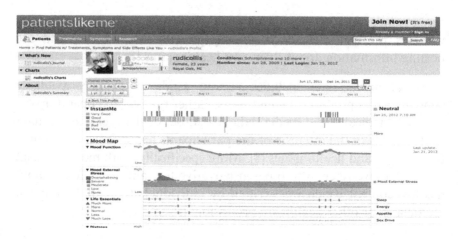

Figure 4.1 Screenshot of PatientsLikeMe. Data is directly plotted to graphs showing changes in a person's health status over time. Source: http://trendreport.betterplace-lab.org/case/patientslikeme.

Content sharing media allow anyone to upload content such as videos or pictures to be shared with everyone or with a restrictive community of users. *Collaborating knowledge sharing social media* such as forums enable users to ask questions and wait for answers coming from different users. In crisis situations such tools can be exploited for information exchange including images and videos.

Weblogs or blogs are similar to paper-based diaries that are normally kept by individuals and shared with others. Similar to a paper-based diary, the authors describe their personal opinions, impressions, and feelings. Online reviews of medical products are an additional source of information regarding the efficacy and adverse effects of drugs and medical devices. A microblog is a blogging platform where the amount of information that can be shared per author is very short. The most common examples of microblogs are Twitter and Tumblr. Twitter's limit is set such that a standard text message, which is limited at 160 characters, can include one entire tweet plus address information. Besides individuals, organizations are tweeting. For example, the U.S. Center of Disease Prevention and Control (https://twitter.com/cdcflu) and the organization Medécins sans Frontiers (https://twitter.com/MSF) are tweeting updates on disease activity (Table 4.1). Systems that collect information on disease activity have channels where detected activities are posted [e.g., HealthMap (https://twitter.com/healthmap), ProMED-mail (https://twitter.com/ProMED_mail)]. Furthermore, vaccination or disease prevention campaigns are supported by information distribution through

Table 4.1 Example of Tweets on Swine Flu, Posted on March 15, 2010

[WEBMD] H1N1 Swine Flu Still Smoldering in U.S.—It's no wildfire, but H1N1 swine flu continues to smolder in the... http://ow.ly/16PbAi
[FluGov] March 12—WHO Updates International H1N1 Flu Situation. http://ow.ly/16OwoZ
[CDC] UPDATE: CDC Estimates of 2009 H1N1 Influenza Cases, Hospitalizations and Deaths in the United States, April 2... http://ow.ly/16Owp0

Table 4.2 Types of Social Media and Their Use for Health Risk and Crisis Communication and Digital Epidemiology

Type of Social Media	Example	Use for Risk and Crisis Communication and for Digital Epidemiology
Social network	PatientsLikeMe, Facebook	Coordination among emergency services and volunteers, share information inside a community, swift update on emergency situation
Content sharing	YouTube	Situational awareness in real time through exchange of pictures and videos, launch vaccination or disease prevention campaigns
	FlickR	
Collaborating knowledge sharing social media	Wikis	Situational awareness, dialog between victims and emergency services
	Forums	
	Message boards	
Blogging/microblogging	Blogger	Convey recommendations, warnings, share facts and rumors, source for mapping emergency information
	Twitter	
	WordPress	

twitter or other blogging platforms. Twitter has been proven a frequently updated data source and many technologies analyze twitter messages for the purpose of detecting public health threats [5].

In summary, these different social media are a source of patient-collected clinical values (e.g., blood pressure, pulse, weight), individual judgements on symptoms or efficacy of drugs and treatments, and feelings and sentiments reflecting the health status. Table 4.2 summarizes potential use cases for the four social media types in digital epidemiology and crisis communication.

4.2.2 An Example of Twitter as Data Source

Twitter messages have a common format: [username] [text] [date time client]. The linguistic variety goes from complete sentences to listing of keywords. Hashtags, i.e., terms that are combined with a hash (e.g., #flu) denote specific topics and are primarily utilized by experienced users.

Table 4.3 Categorization of Tweets		
Content	**Description**	**Example Tweets**
1. Resources	Resource tweets contain news, updates, or information about diseases or outbreaks. The title of the linked article might be mentioned	#schweinegrippe Neue Schweinegrippefälle in Europa: Kulmbacher Gesundheitsamt warnt vor P... http://tinyurl.com/36o4nqh #influenza #h1n1 (Translation: #swineflu New cases of swine flu in Europe: Health ministry of Kulmbach warns...)
2. Personal opinion with linking	Twitter users post their opinion on a disease, virus, symptom	Schweinegrippe : D http://yfrog.com/h3zuhtj (Translation: swineflu :D)
3. Personal opinion and information	Twitter users post only their personal feelings or health status	Hallo Freitag—Hallo Erkältung (Translation: Hello Friday—hello cold)
4. Marketing	Tweets contain an advertisement for a disease-related product or service	It's National Influenza Vaccination Week. Get vaccinated to fight flu!
5. Spam	Tweets are unrelated to diseases or symptoms, but contain mentions of diseases or symptoms	"Das soll gegen Erkältung helfen!" http://twitpic.com/3fl9q0 (via @haraldmeyer) (Translation: This is expected to help against a cold)

Referring to the study from Chew and Eysenbach [9] we categorize tweets according to their contents (Table 4.3). In more detail, Twitter messages can:

• provide information,
• express opinions,
• report personal issues.

Information or resources can be provided by authorities, individuals, news agencies, or health professions (Table 4.4). If information is provided, the authority of that information can normally not be determined, so it might be unverified information. Opinions are often expressed with humor or sarcasm and may be highly contradictive in the emotions that are expressed. Consider for example the tweet: "I feel so sick. I have Bieber fever:-)." On the one hand, it reports about the sickness which is rather negative. On the other hand, there is the smiley, which denotes that there is no serious illness, but only "Bieber fever," which is not really a disease, but rather it is related to the pop star Justin Bieber.

Tweets that contain mentions of symptoms or diseases can be further distinguished based on their content in informing about the

Table 4.4 Author Groups and Their Content Distributions in Twitter

Author Group	Provided Content
Individual, nonhealth professional	Personal symptoms
	Observation of symptoms in others
	Reposting information from others including news agencies or health institutions
Individual, health professional	Observation of symptoms in others
	Reposting information from others including news agencies or health institutions
News agencies	Mainly official information
Health institutions	Official information, warnings

health status of the (1) author of the tweet, (2) a friend of the author, or (3) a prominent person. Rarely, they are reporting about health status of animals. Further, personal tweets and resources are reporting about general health information or health education, official information or advices from travel medicine. Characteristically, tweeters are using short sentences (e.g., *I have fever*) or just keywords (e.g., *fever, cough, headache*). Abbreviations are widespread and sometimes difficult to understand due to a lack of context.

When looking into the content of tweets, we can recognize that the various user groups provide different types of content through the twitter channel (Table 4.4).

Challenges for automatic processing of tweets are related to the unstructured nature of the data (free text) and layman language which hamper the connection to clinical terminology. Another issue is the volume of data that is available as well as its reliability. Associated with the reliability of data is the difficulty in interpretation and evaluation of the data. The quality of the data provided through social media tools is unknown. It can be comprehensive and helpful, and also misleading or wrong. Different terminologies and semantics complicate an automatic analysis and interpretation. Subjective information needs to be interpreted, weighted, and linked to objective clinical parameters.

4.3 TECHNOLOGIES

Practice showed that a substantial amount of initial outbreak reports come from unofficial informal sources distributed through social

Table 4.5 Summary of Internet-Based Outbreak Detection Systems			
System Name	**Data Sources**	**Technology**	**Description**
GPHIN	Online news, websites	Taxonomy, translation, categorization	Disease outbreak warning tool
BioCaster	RSS feeds, Google news, WHO, ProMED-mail, European Media Monitor	BioCaster Ontology	Online media data monitoring system
HealthMap	Baidu News, SOSO Info, Google News, Moreover, GeoSentinel, WHO, ProMED-mail	Google Maps, Google Translate, Fisher–Robinson Bayesian filtering	Global disease alert map
EpiSpider	ProMED-Mail, medical websites	OpenCalais, UMLS	Global disease alert map
MedISys	RSS feeds from news, blogs, and official sources	Lexicon, keyword list	Monitoring tool for infectious diseases and chemical, biological, radiological, and nuclear hazards, statistics
Google Trends	Google search term statistics	Word statistics	Trend monitoring system can be used for flu and also other medical conditions

media platforms. These messages need verification. Thus, there is a need to support health officials and epidemiologists to identify relevant information related to disease outbreaks.

4.3.1 Social Media for Surveillance and Detection of Outbreaks

Surveillance and outbreak detection tools use different sources of web data that is checked for disease names, mentions of symptoms or other features enabling an identification of relevant postings or web content [10]. Some systems rely upon keyword lists, others on ontologies. Most of them are processing content in different languages, focusing on global disease surveillance. Interestingly, the systems are based on different knowledge resources ranging from keyword lists to taxonomies and ontologies. Even the Unified Medical Language System [11] is exploited as knowledge source by one system. Some examples of such systems are described in Table 4.5.

4.3.1.1 Global Public Health Intelligence Network

The Global Public Health Intelligence Network (GPHIN [12]) is an electronic public health early warning system developed by Canada's Public Health Agency which is part of the World Health Organization's (WHO) Global Outbreak and Alert Response Network (GOARN).

This system monitors internet media, such as news and websites, in nine languages in order to help detect and report potential disease outbreaks or other health threats globally. From 2002 to 2003, this surveillance system was able to detect the Severe Acute Respiratory Syndrome (SARS) outbreak.

More specifically, GPHIN is a secure, web-based restricted access system for outbreak alert that deals with news information about public health events. In contrast to traditional surveillance systems that rely on subscriber input, GPHIN gathers information on disease outbreaks and other public health events by monitoring global media sources on a 24/7 basis. GPHIN's two main sources of outbreak information are the global news services Factiva and Al Bawaba in Arabic language. These services operate as news aggregators that provide multiple sources of information through a single access point. Factiva, for example, aggregates news information from nearly 9000 sources in 22 languages.

GPHIN works in five main steps: From collected data duplicates are eliminated; texts are translated and metadata is inserted using a taxonomy (e.g., mentions of "SARS" or "H1N1" are recognized as "human diseases"). Then, the data are categorized and a relevance score is determined. All data considered relevant is published and available for a manual analysis triage.

4.3.1.2 BioCaster
BioCaster [13] is a project aimed at providing advanced search and analysis of internet news and research literature for public health workers, clinicians, and researchers interested in communicable diseases. The system monitors many hundreds of internet newsfeeds simultaneously to detect and track infectious disease outbreaks (Fig. 4.2).

More specifically, the system continuously analyzes documents reported from over 1700 RSS feeds including Google News, WHO, ProMED-mail, and the European Media Monitor. The extracted portions of text are classified for topical relevance and plot onto a Google map using geoinformation. The system works in four main steps: topic classification, named entity recognition (NER), disease/location detection, and event recognition.

In more detail, the BioCaster system is equipped with text mining technology which continuously scans hundreds of RSS newsfeeds. The

Figure 4.2 BioCaster user interface.

text mining system has a detailed knowledge about the important concepts such as diseases, pathogens, symptoms, people, places, and drugs. This allows to semantically index relevant parts of news articles, enabling precise access to information. The knowledge underlying the text mining algorithm comes from annotated text collections, gazetteer lists of nomenclature, and the BioCaster ontology. The BioCaster system is no longer accessible online.

4.3.1.3 HealthMap

HealthMap [14], available at http://www.healthmap.org/en/, is a platform developed by a team of researchers, epidemiologists, and software developers at Boston Children's Hospital founded in 2006. The system exploits online informal sources for disease outbreak monitoring and real-time surveillance of emerging public health dangers.

Similar to the systems described before, HealthMap collects data from different data sources, including online news, eyewitness reports, expert-curated discussions, and validated official reports. Via an automatic system, which is being updated 24 hours per day, this system monitors, organizes, integrates, filters, visualizes, and disseminates online information about emerging diseases in nine distinct languages, facilitating early detection of public health threats. Collected data is processed by means of automated filtering, and visualization of reports

Figure 4.3 Health Map alert map. Source: http://www.healthmap.org/en/.

through the utilization of automated text processing algorithms that classify alerts by location and disease. A mean against information overload, the articles are further categorized for improved filtering. The additional categories include breaking news (e.g., of a newly discovered outbreak), warning, follow-up, background/context, and not disease related (Fig. 4.3).

4.3.1.4 EpiSpider

EpiSpider [15] was initially developed to serve as a visualization supplement to the ProMED-mail reports (http://www.promedmail.org/), i.e., ProMED-mail reports were analyzed with respect to topic intensity and displayed on a map. In addition to ProMED data, EpiSpider collects information from Google, Humanitarian News, Twitter, WHO, and Daylife (cloud-based media service) and processes the data with natural language processing used to transform free text into structured information. EpiSPIDER began outsourcing some of its preprocessing and natural language processing tasks to external service providers such as OpenCalais (www.opencalais.com) and the Unified Medical Language System (UMLS) web service for concept annotation. This action has enabled the screening of noncurated news sources

Figure 4.4 MedISys user interface. Source: http://medusa.jrc.it/.

as well. However, it scans articles only in English. EpiSPIDER has a timeline visualization to help users to order events in time, and a word cloud that helps users to get a sense of what topics are making headlines. Location names in reports are recognized and georeference using the georeferencing services of Yahoo Maps, Google Maps, and Geonames.

4.3.1.5 MedISys

MedISys [16] is an internet monitoring and analysis system that identifies potential threats to the public health (Fig. 4.4). Collected articles are grouped by disease or disease type; location names are identified. The system is based on a list of sources, including official channels, blogs, and online news. It analyzes the texts using keyword lists and identifies topics that are focused by many reports at the same time span. MedISys (http://medusa.jrc.it/) covers global health issues including multiple diseases and multiple locations.

4.3.1.6 Google Trends

Google trends (https://www.google.com/trends/) are another emerging tool for the detection of outbreaks. It uses search query data, i.e., frequency of search terms, and plots them over time, allowing for the

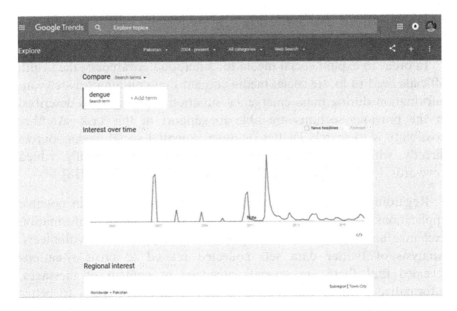

Figure 4.5 Google trends showing dengue searches from 2004 to present in Pakistan. Source: https://www.google.com/trends/.

recognition of potential disease activity when considering a disease-related term. As a result, a time series of the searches in a period of time is provided giving an idea how the search trend changed during defined time period (Fig. 4.5). This in turn gives an idea of possible disease activity which might have occurred. Indeed, studies have shown a correlation between search behavior and flu activity [17].

4.3.2 Social Media for Health Crisis Communication

Emergency management and crisis communications have become more participatory. Through Social Media channels information on disease activity or natural disasters can be distributed quickly to large groups of individuals. The OECD claims: "Social Media can enhance risk and crisis communication in several ways: (1) they are collaborative and participatory, and thus can improve situation awareness, (2) they are decentralized, thus, information can circulate quickly, and (3) they are geographically traceable and thus allow for the monitoring of a crisis" [5].

No specific system is required for social media-based health crisis communication—the use of the existing social media tools such as social networks (e.g., Facebook) or microblogging systems is sufficient for crisis communication.

In order to exploit social media for situational awareness, the health officials need to locate social media content that contain crisis-relevant information during mass emergency situations. The systems described in the previous sections are able to support in this task. Another possibility is to search Twitter or other potential social media sources directly which can be realized by conventional manually edited keywords, location-based searches, or relying upon lexicons [18].

Regarding circulation of information, there are four main possible applications: distribution of information to the public, information exchange among staff and volunteers, and acquisition of volunteers. Analysis of twitter data sets collected related to crisis situations revealed that there are several categories of content of messages. Informative postings can contribute contextual information to better understand the situation. They include status messages of users, explanations of particular problems, and precise data of the crisis, e.g., number of victims. The content can be predicting or forecasting, instrumental or conformational [19].

CDC officials use various social media channels to inform the public and provide health and safety information. The CDC Emergency feed (https://twitter.com/CDCemergency) is the official feed of CDC's Office of Public Health Preparedness and Response. It provides latest information on emergencies, preparedness tips, and real-time updates and health alerts for the public during an emergency. Additionally, email updates are provided with information on recent outbreaks and incidents, radiation emergencies, or public health matters. More specifically, the E-Mail alerts are generated when new information on the corresponding CDC website is available.

During the Ebola outbreak in 2014, social media supported the communication between healthcare providers, local and national health authorities, and international health agencies. Furthermore, social media has been proven to be able to replace traditional communication systems during crisis situations [5]. During the 2010 earthquake in Haiti, traditional communication systems were down. People started to use social media quickly as communication channel [20].

Additionally, social media can be used to indicate willingness to help in the event of an emergency and thus may help in mobilizing volunteers [21]. For example, indicating in the "status" of the personal Facebook profile, the availability and skills for both professionals and volunteers could be a way for public authorities to know in real time whom to mobilize in a given area of disaster.

4.4 THE M-Eco System and Lessons Learnt

The EU-funded project M-Eco: Medical Ecosystem was conducted between 2010 and 2012 with seven project partners from Austria, Italy, Germany, Czech Republic, and Denmark, including the German health organization Robert Koch Institute and with support of representatives of various health organizations including the World Health Organisation, European Centre of Disease Prevention and control, and Institute de Veille Sanitaire. In this section, we briefly summarize the architecture of the M-Eco system and its functionalities, as well as report on experiences in its evaluation and testing. The M-Eco system could so far not been established into regular use by health organizations due to missing personnel resources from implementation. More details about the technology and studies can be found in papers by Denecke et al. [22] and Velasco et al. [10].

4.4.1 Overview

The M-Eco system exploited data from multiple sources for public health monitoring purposes. The system:

- monitored social media, TV, radio, and online news,
- aggregated the contents into signals,
- visualized the signals using geographic maps, time series, and tag clouds,
- allowed searching and filtering signals along various criteria (location, time, medical condition).

The system was intended to support in health monitoring during mass gathering events in a cross-country setting and in health monitoring on a national level. Signals pointed the user to relevant information and their sources which allowed to analyze its relevance and need for interaction through health officials. Automatically generated time series supported the monitoring of disease activity over a longer time period. Tag clouds summarized the related information in a visual manner and supported

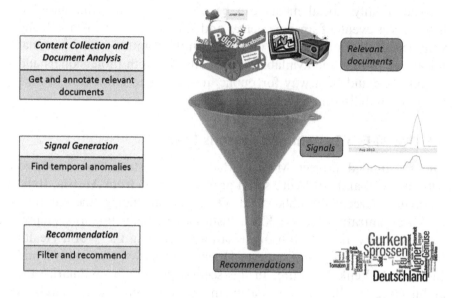

Figure 4.6 The M-Eco approach in a nutshell.

navigation through signals. The plotting of signals to geographic maps allowed to localize disease outbreaks. M-Eco offered (1) additional information through social media monitoring, (2) perception of recommendation and user behavior, and (3) visualization and support for risk assessment.

4.4.2 Architecture

To realize these steps, the M-Eco system consisted of a set of web services that cover four areas depicted in Fig. 4.6. These were (1) content collection, (2) signal generation, (3) user modeling and recommendation, as well as (4) visualization in a user interface. The services worked in a pipeline fashion and were triggered automatically four times a day.

Content collection and document analysis component

The information database of the system was filled continuously by the content collector and document analysis component. It collected data from various sources by means of web crawling and streaming APIs (e.g., the Twitter API), and made them accessible to other components. The collection focused on broadcast news from TV and radio, news data from MedISys [16], and social media content from blogs, forums, and Twitter. The TV and radio data was collected via satellite and transcribed to written text by SAILs Media

Mining Indexing System [23]. About 1300 names of symptoms and diseases were used as keywords for collecting data extended by existing language resources such as WordNet, GermaNet, or the OpenOffice thesaurus. The data was tokenized and part-of-speech tagged by the Tree Tagger and parsed by the Stanford Parser. All texts were also semantically annotated with geo tags, disease or symptom tag and temporal expressions as well as with information on the affected organism. In the following, the term "text" is used to refer to some piece of text which can be, e.g., a tweet, a blog posting, or even a transcript of a TV or radio transmission.

Signal generation component
The event detection and signal generation component exploited the annotated texts provided by the content collection and document analysis component to generate signals. A signal was a hint to some anomalous event. The component produced signals with associated information on the disease or symptom the signal is referring to and a location that has been extracted for that signal. In a first step, sentences were classified as relevant or irrelevant by the method presented before. For all relevant sentences, entity pairs (location, disease) were exploited to produce time series for each entity pair occurring in sentences of texts published within 1 week.

The time series provided the input for statistical methods for signal generation, CUSUM and Farrington. These two statistical methods had originally been developed for indicator-based surveillance [24]. Cumulative sum or CUSUM methods focused on several consecutive periods and sum up the aberrations in one particular direction. The Farrington approach fitted a regression model to the data over several years, allowing for a secular trend. Outbreaks in the past were automatically identified and removed, and the statistical distribution fitted either to rare counts or to frequent counts.

The user interface allowed to select the algorithm used to calculate signals. Between 0 and 50 signals were generated by this procedure every night. The exact number depended on several variables or factors that influence the generation of signals such as the type of considered data (e.g., Twitter's update frequency was much higher than of a blog).

Recommendation component
The recommendation component got as input the generated signals and either selected those that were of interest for a user according to his profile or ranked the signals appropriately. The component

also supported users with personalized presentation options (e.g., tag clouds, list of recommendations) that were visualized in the user interface. In this way, information or alerts were filtered before being presented to a user, which in turn reduced information overload. The recommendation component required a user profile that consisted of information on user behavior from interactions with the system (e.g., ratings, tags, search terms).

The personalization and recommendation of signals mainly relied upon the tagging behavior of a user. Tags were potential indicators of user preference. For recommending items to the user, tags assigned by him to his texts of interest were compared against the tags assigned to candidate and unknown texts. In order to help users navigating through a vast collection of texts and finding new items, a tag cloud component provided a visual representation of texts. Besides indexing texts in the corpus, each tag helped users to find new related information of interest.

User interface and visualization

The user interface allowed a user to search for disease names or symptoms and to assess the related signal information by means of a geographic map, a tag cloud, or a timeline. The geographic map plotted the signals to a map. It enabled users to select specifically signals related to locations that are of interest to them. The timeline showed the text volume referring to a specific disease or symptom (or the corresponding signal, respectively) over time. This allowed users to learn about the progress of a disease outbreak as reflected in social media and also about seasonal differences. The tag clouds provided a quick overview on the content of the texts associated with a signal. They enabled the user to quickly decide about the relevance of a signal. Access to the original sources that contributed to the signal generation was provided, as well as filtering capabilities (e.g., selecting a time span). Furthermore, user feedback options were included into the user interface. With "Thumbs up—thumbs down" and a rating scale for signals, users could judge the relevancy of the presented signal. This information was fed back to the recommendation process and considered for ranking and filtering.

4.4.3 Lessons Learnt

The M-Eco system results were analyzed in several studies [10,22]. They revealed characteristics of social media that are relevant for disease surveillance. First, the texts that contributed to signals rated as relevant by the epidemiologist were often linked to media reports or so-called secondary reports. This suggests that there might be a trend in social media whereby users tend to write less often about their personal specific symptoms, but most often forward information from reliable sources such as news sites or prevention efforts from authorities.

Second, most signals were generated from Twitter data. The volume of relevant Twitter data that was processed by the system was much higher than that from any other source considered as input. Beyond, it is unclear who is providing relevant health information via social media, which age groups, personal background of persons might play a role, geographic coverage, etc. This means that relevant information from segments of the populations are not coming through these channels. Another challenge is the quality of content collected from social media and the difficulty to automatically decide whether it is a real outbreak or not. Many of the social media texts present vague reports of illnesses. It is difficult to judge the seriousness of the reported information.

In contrast to initial expectations, the signals were not generated from clustered reports on personally reported symptoms, but on news reports that were fed into social media, and replicated or forwarded by interested users. Therefore, M-Eco was not the first instance to detect the public health event, because there were local actors who had already detected and reported the event. But, M-Eco brought such reports quickly to a broader attention. It was not possible to present an example where M-Eco was the first to detect an outbreak by a clustering of social media contributions with similar symptoms in space and time, and where the outbreak was afterwards confirmed by the traditional notification system. Another lesson learnt is not to underestimate the legal and ethical issues related to IT solutions for digital epidemiology. We discuss relevant issues in Section 4.6.

4.5 SWOT ANALYSIS

By means of an analysis of the strengths, weaknesses, opportunities, and challenges, we are studying the potential of social media for disease surveillance and crisis communication and prevention.

The objective is to identify future perspectives and open issues to make social media a useful tool in this context. Strengths and weaknesses are mainly internal factors while opportunities and threats

Table 4.6 SWOT Analysis Results		
	Positive	Negative
Internal	Strengths	Weaknesses
	• Easy to use • Availability without any installation • Low costs • Fast and timely • Enables direct contact to public and volunteers • Enables a large coverage through a broad distribution • Can be used by professionals and laymen • Reception of outbreaks that would be otherwise recognized too late	• Information overload • Entirely subjective information, unverifiable • Difficult to assess and check validity • Unauthorized information • Unknown source of information; even location can be unknown • Wrong or misleading information can be posted
External	Opportunities	Threats
	• Enables interactions among health organizations and volunteers • Engagement of the population in prevention • Increased availability of internet in rural areas • Large distribution of smart and mobile phones also in low income countries • Health officials use social media for global disease surveillance and prevention campaigns	• Content is unbalanced with respect to information provider (younger persons) • Risk of manipulation/spam • Availability of internet access is a must—difficult in rural areas • Data privacy: people are sharing personal information in the web, but when coming aware of privacy and security issues they might stop sharing • Reliability of information • Technology needs to be able to resolve ambiguities and filter out irrelevant information • New standards or laws could forbid the use of social media for monitoring purposes • Ethical issues might hamper the use

generally relate to external factors. For the analysis, we collected answers to the following questions (Table 4.6):

Strength
- What advantages does social media have for digital epidemiology?

Weaknesses
- What can be improved in current tools and their usage? What should be avoided?

Opportunities
- What interesting trends and opportunities can be identified?

Threats
- What obstacles can be faced when implementing using social media for disease surveillance and crisis communication?
- Do quality standards or ethical and legal aspects threat the use in practice?

It can be seen that there are many positive aspects supporting the use of social media for disease surveillance and crisis communication. More timely information and networking among the "helping hands" could become the most driving factor. Challenges are related to the increased information overload, including the amount of unauthorized information which is information not officially confirmed by health officials, which might be addressed by technology, e.g., by including sophisticated filtering algorithms to prefilter the information before showing to the user. A challenge is the high risk of manipulation, in particular the risk of analyzing postings containing wrong or misleading information. The usefulness of social media for disease surveillance depends clearly on the willingness of people to share (correct) information online and to use online tools. In particular, we will need in future ethical and legal standards to ensure that people will continuously use these tools for reporting on disease activity. In the next section, unintended consequences, in particular ethical and legal issues will be discussed in more detail.

4.6 UNINTENDED CONSEQUENCES OF SOCIAL MEDIA USAGE IN PUBLIC HEALTH

Even though useful, social media usage for prevention and detection of epidemics provides some unwanted or unintended consequences with regards to technical, functional, and formal issues.

Formal problems include quality and reliability of content, payment models, as well as ethical and legal issues. The latter are related to the usage of data posted through social media tools for research or epidemiological purposes. In this context, it is important to clarify responsibilities. Imagine a health status monitoring tool that identifies a group of sick persons by analyzing social media conversation. In which manner should a health organization react that becomes aware of this conversation? The current interaction or reaction processes are often not foreseeing a reaction of the national health organizations, but on a local level. This means, processes need to be adapted when considering social media as source of information or for crisis communication. These and similar questions need to be answered before such applications go online. When using social media or online traces for surveillance purposes, the right for individual self-determination—what happens with my data—is weighted in crisis situations against the wellbeing of the society. The objective of data collection and analysis needs to be specified, i.e., whether the data is analyzed and collected for treatment, care, prevention, or crisis management.

In both cases, disease surveillance and crisis communication, it needs to be ensured that the technology is robust against errors and abuse in order not to overload the health officials with information, but also to prevent population groups from social stigmatizations and prejudices due to false alarms. Corresponding measures (e.g., integrating spam detection methods) should be implemented or personnel needs to be aware of misleading information.

Technical and functional challenges are related to the data volume and an increased risk of generating false alarms. Even though social media data provide a new source of information to hint to public health threats, their analysis and interpretation is challenging. Language is ambiguous and automatic interpretation becomes difficult when symptoms are used in different contexts than expected (e.g., "football fever" could produce an alert since the keyword « fever » is used). Intended as support for epidemiologists and healthcare workers, there is a risk of an additional workload due to large numbers of such false alarms. Comprehensive filtering algorithms need to be established keeping a good balance between sensitivity and specificity of generated alarms. Another option is to use social media tools for active reporting on disease activity by the population. However, even this method is prone to errors when misleading information is posted.

Another issue is the quality and reliability of data as well as localizing the outbreak. As reported by Goff et al. [25], sometimes misinformation regarding infectious diseases is disseminated through Twitter. An additional limitation is that majority of users of social media are younger people from developed countries. This makes social media-based information biased spreading misleading information as reported by Paul et al. [26]. Dyar et al. [27] found out that Twitter leads to activation of certain searches and sharing of information about outbreaks globally rather than locally. This can become misleading as the outbreak can occur in a certain part of the world while information is being shared to other parts.

Maintaining traditional media and reporting in the crisis communication strategies and for disease surveillance are relevant to ensure inclusion of all segments of the population. Beyond, there are other measures to be taken to make the best use out of social media in digital epidemiology and crisis communication. When developing a concrete social media application in epidemiology and for the detection of epidemics, it is crucial to determine the scope of the system under development, i.e., it needs to be clarified which users are involved, which application area is concerned and on which dimension it is operated. Questions include:

- Who is affected by the analysis and application of medical social media data and how should they be affected by it?
- Who is compelled to act on the new knowledge?
- What action is appropriate based on the information learned as a result of the analysis?
- Who is responsible when a predictive analysis is incorrect?

Answering those questions before implementing a system in practice and even addressing these questions in the development phase will help in producing useful applications, limiting the risks of social media usage for prevention of epidemics. However, there is still a need for guidelines, standard operating procedures, and best practices in digital epidemiology to ensure that harm is prevented.

REFERENCES

[1] Paquet C, Coulombier D, Kaier R, Ciotti M. Epidemic intelligence: a new framework for strengthening disease surveillance in Europe. Euro Surveill 2006;11(12). Available from: <http://www.eurosurveillance.org/ViewArticle.aspx?ArticleId=665>.

[2] WHO: Health topics. Available from: <http://www.who.int/topics/disease_outbreaks/en/> [last accessed 12.11.15].

[3] Funk S, Gilad E, Watkins C, Jansen VAA. The spread of awareness and its impact on epidemic outbreaks. Proc Natl Acad Sci 2009.

[4] Salathé M, Bengtsson L, Bodnar TJ, Brewer DD, Brownstein JS, Buckee C, et al. Digital epidemiology. PLoS Comput Biol 2012;8(7):e1002616. Available from: <http://dx.doi.org/10.1371/journal.pcbi.1002616>.

[5] OECD. The changing face of strategic crisis management. Paris: OECD Publishing; 2015. Available from: http://dx.doi.org/10.1787/9789264249127-en.

[6] Whitney Holmes. Crisis communications and social media: advantages, disadvantages and best practices. CCI Symposium. Available from: <http://trace.tennessee.edu/cgi/viewcontent.cgi?article=1003&context=ccisymposium> [last accessed 15.02.16].

[7] Smith BG. Socially distributing public relations: Twitter, Haiti, and interactivity in social media. Public Relations Rev 2010;36(4):329–35.

[8] Denecke K. Health web science. Social media data for healthcare. Springer International Publishing, Switzerland; 2015.

[9] Chew C, Eysenbach G. Pandemics in the age of Twitter: Content Analysis of Tweets during the 2009 H1N1 Outbreak. PLoS One 2010;5(11):e14118. Available from: http://dx.doi.org/10.1371/journal.pone.0014118.

[10] Velasco E, Agheneza T, Denecke K, Kirchner G, Eckmanns T. Social media and internet-based data in global systems for public health surveillance: a systematic review. Milbank Q 2014;92(1):7–33.

[11] Lindberg DA, Humphreys BL, McCray AT. The Unified Medical Language System. National Library of Medicine, Bethesda, MD. Methods Inf Med 1993;32(4):281–91.

[12] Mykhalovskiy E, Weir L. The global public health intelligence network and early warning outbreak detection. Can J Public Health 2006;97(1).

[13] Collier N, Doan S, Kawazoe A, Goodwin RM, Conway Mike, Tateno Y, et al. Biocaster: detecting public health rumors with a web-based text mining system. Bioinformatics 2008;24(24):2940–1.

[14] Brownstein JS, Freifeld CC, Reis BY, Mandl KD. Surveillance sans frontieres: internet-based emerging infectious disease intelligence and the healthmap project. PLoS Med 2008;5(7):e151.

[15] Keller M, Blench M, Tolentino H, Freifeld CC, Mandl KD, Mawudeku A, et al. Use of unstructured event-based reports for global infectious disease surveillance. Emerg Infect Dis 2009;15(5):689.

[16] Linge JP, Steinberger R, Fuart F, Bucci S, Belyaeva J, Gemo M, et al. MedISys: medical information system. Advanced ICTs for disaster management and threat detection: collaborative and distributed frameworks. IGI Global Press; 2010.

[17] Cook S, Conrad C, Fowlkes AL, Mohebbi MH. Assessing Google Flu trends performance in the United States during the 2009 Influenza Virus A (H1N1) pandemic. PLoS One 2011;6(8):e23610. Available from: http://dx.doi.org/10.1371/journal.pone.0023610.

[18] Olteanu A, Castillo C, Diaz F, Vieweg S. CrisisLex: a Lexicon for collecting and filtering microblogged communications in crises. Proceedings of the AAAI conference on weblogs and social media (ICWSM'14). Ann Arbor, MI: AAAI Press; 2014.

[19] Sreenivasan ND, Lee CS, Goh DH-L. Tweet me home: exploring information use on Twitter in crisis situations. Online Communities and Social Computing: 4th international conference, OCSC'11, Held as Part of HCI International 2011, Orlando, FL, USA, July 9–14, 2011. Proceedings. Springer Berlin/Heidelberg. p. 120–9.

[20] Cambié S. Medecins sans frontieres: social media lessons from the Haiti crisis. Available from: <https://www.simply-communicate.com/case-studies/company-profile/medecins-sans-frontieres-social-media-lessons-haiti-crisis> [last accessed 20.12.15].

[21] Wendling C, Radisch J, Jacobzone S. The use of social media in risk and crisis communication. OECD Working Papers on Public Governance No. 24, OECD Publishing; 2013.

[22] Denecke K, Krieck M, Otrusina L, Smrz P, Dolog P, Nejdl W, et al. How to exploit twitter for public health monitoring? Methods Inf Med 2013;52(4):326–39.

[23] Backfried G, Schmidt C, Pfeiffer M, Quirchmayr G, Glanzer M, Rainer K. Open source intelligence for disaster management. intelligence and security informatics conference (eisic); 2012. p. 254–58.

[24] Hoehle M. Surveillance: an r package for the surveillance of infectious diseases. Comput Stat 2007;22(4):571–82.

[25] Goff DA, Kullar R, Newland JG. Review of twitter for infectious diseases clinicians: useful or a waste of time? Clin Inf Dis 2015;60(10):1533–40.

[26] Paul MJ, Dredze M. You are what you Tweet: analyzing Twitter for public health. In: ICWSM; 2011. p. 265–272.

[27] Dyar OJ, Castro-Sánchez E, Holmes AH. What makes people talk about antibiotics on social media? A retrospective analysis of Twitter use. J Antimicrob Chemother 2014; dku165.

CHAPTER 5

Big Data For Health Through Social Media

M.A. Mayer[1], L. Fernández-Luque[2], and A. Leis[1]
[1]Universitat Pompeu Fabra (UPF), Barcelona, Spain [2]Hamad Bin Khalifa University, Qatar Foundation, Doha, Qatar

- Introduction about general concepts related to Big Data
- Big Data methodologies applied to Social Media: opportunities and examples
- Challenges of Big Data applied to Social Media: technical, ethical and legal
- Conclusions
- References

5.1 INTRODUCTION

"Big Data" has become a buzzword and is one of the trending topics in several fields such as information management, data analytics, marketing, and healthcare. Although there are several definitions of Big Data, in general, the concept describes a new generation of technologies and architectures that allow us to extract value from massive data volumes and types by enabling high-velocity capture, discovery, and analysis of distributed data. Big Data is also defined by the so-called four V's: *volume* and a new level of complexity of data and new distributed approaches for managing the information; *velocity* of collecting, storing, processing, and analyzing data; *variety* in relation with a large number of different types of data (structured, unstructured, and semistructured); and *veracity* or "data assurance" about data quality, integrity, and credibility [1—4]. The International Medical Informatics Association (IMIA) working group on "Data Mining and Big Data Analytics" defined Big Data as data whose scale, diversity, and complexity require new architecture, techniques, algorithms, and analytics to manage it and extract value and hidden knowledge from it [5]. However it is critical to

Participatory Health Through Social Media. DOI: http://dx.doi.org/10.1016/B978-0-12-809269-9.00005-0

consider that diverse aspects related to the Big Data concept and its application can vary by domain, depending on what kinds of software tools are available in each case and what size and types of datasets are more common in a particular field; each type of Big Data requires the use of some particular analysis methods and tools [6,7].

The impact of the new emerging scenarios related to the management of Big Data is so significant that several institutions and governments set up particular and strategic initiatives to support the development of technologies to manage the huge amount of information accessible in a more efficient and productive manner. For instance, in 2012 the Obama Administration announced the "Big Data Research and Development Initiative" with the aim of advancing, developing, and using Big Data technologies applied to several areas such as health, defense, energy, and climate change [8]. Likewise, in 2014 the European Commission outlined a similar strategy on Big Data in order to develop actions in various fields such as health, food security, climate and energy, intelligent transport systems, and smart cities [9]. In addition, specific initiatives such as the Big Data Value Association (BDVA) has been created in order to boost Big Data research through the establishment of a Public–Private Partnership (PPP) to implement a strategic roadmap for research, technological development, and innovation in the Big Data Value and other Information and Communication Technologies areas [10].

To understand better some facts related to Big Data, it is interesting to mention that in the past decade, large genomics Big Datasets have been created and the different "omics" methodologies such as metabolomics, proteomics, transcriptomics, and epigenomics that have been reached their maturity; for this reason the development of several specific biomedical informatics tools to exploit and mine this type of information can be applied in other domains that requires similar analytic efforts [7,11,12]. On the other hand, the reuse of clinical information for research and the integration of clinical, medical, and public health electronic systems is becoming a reality. Through the integration of the diverse sources of clinical information, which at the current time often reside in silos and separate repositories, it is possible to analyze Big Data helping researchers to find relationships among variables, not detectable in the past. Furthermore, new knowledge about the effectiveness and side effects of treatments can be generated reshaping biomedical research [13–17].

In this context, in step with the explosive growth of online services and platforms on the Internet, more and more organizations, professionals, and scientific institutions are seeing the need to make the most of other sources of health information such as the Social Media (SM) platforms through the use of Big Data tools and analytics. Although SM used as a source of health information is generally of much lower quality, the large amount of data collected through these environments can offer new insights and data that is not present, for instance, in Electronic Health Records (EHRs). SM can be a complementary channel of information to other official means for the health data collection such as the epidemiological surveillance activities and control carried out by health authorities. In recent years it is possible to find a lot of examples that show many applications of the analysis of SM for health purposes such as disease surveillance, health promotion, and public health [18–22] or a parallel source to the official means of pharmacovigilance, as a way to discover new drug side effects associations not described before or new applications for old drugs [23–25]. Several conditions and diseases can be monitored on SM platforms, identifying patterns of behavior that give clues about the personality profile, mental health, and the possibility of success in smoking cessation or eating habits of their users [26].

5.1.1 Big Data Value Chain

The term "Big Data" can be sometimes confusing since the "size" of data is not an objective measurement since, for example, the side of a medical image in terms of data will continuously grow as the data grow. The "big" in terms of describing the computing power needed to process the data is neither a good objective measurement as the computing power is also continuously evolving. Taking into account these issues, sometimes it is best to forget the term of Big Data and focus more on the different processes involved in Big Data (see Fig. 5.1): (1) to extract value for large amounts of heterogeneous data and (2) to create value from the data in order to create services and applications.

As was mentioned earlier, in Europe the BDVA has created a framework where technology around Big Data is classified regarding its position in the value chain. As seen in Fig. 5.2, there are four main areas in the value chain: (1) Data acquisition, (2) Data analysis processing, (3) Data storage curation, and (4) Data visualization and services.

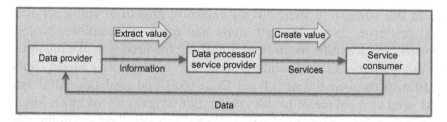

Figure 5.1 Process for the creation of value from Big Data applications [27].

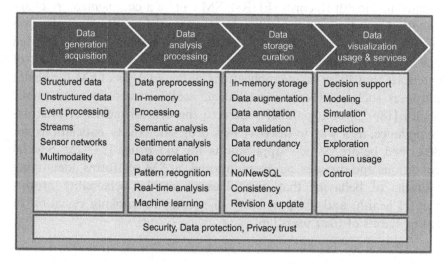

Figure 5.2 Technical elements in the Big Data Value Chain [27].

To extract value from data, the first step is to extract the data. That can be done using sensors (e.g., FitBit, Glucometers) and also integration with EHRs or even free text from online forums and SM platforms. The acquired data normally needs a process of analysis and processing, for example, analysis of web forums from patients will need analysis to extract semantic meaning. The analysis and processing allow us to extract value from the data, but that value needs to be stored in a way that facilitates reuse and ensures quality. Finally the value extracted from the data needs to be used to create and enrich applications and services often requiring new ways to visualize the data.

This chapter is structured as follows. In Section 5.2 we provide an overview of the technical aspects related to the use of Big Data technologies in the health SM domain. Section 5.3 provides some use cases, such as Healthmap. Finally we describe major challenges related to the application of Big Data in SM and conclusions.

5.2 BIG DATA METHODOLOGIES APPLIED TO SM

As described in Section 5.1 we can divide technical aspects into five different areas: (1) Data Generation and Acquisition, (2) Data Analysis and Processing, (3) Data Storage and Curation, (4) Data Visualization Usage and Services, and (5) Security, Data Privacy, and Trust.

5.2.1 Data Generation and Acquisition

Data Generation and Acquisition is a challenging task due to the enormous sources of data available in health SM [28]. Among others, we can consider photos, videos, texts, and metadata about different types of sources.

We need to distinguish between *structured data*, for example, any video published in YouTube has standardized data regarding the video (also called metadata). Similarly, Facebook has defined a structured data structure to share data regarding the social networks (e.g., friends, user profiles, and favorites). Although there are open standards defining those data sources such as OpenSocial [29], most of the SM platform use their own corporate type of structure data sources. The sharing of the data structure together with tools for integrations is crucial for many third-party applications integrated in social network platforms such as Facebook.

Unstructured data such as free text is very common source of data and also a very challenging type since often it requires the use of Natural Language Processing for analysis. An example of unstructured data is the search queries entered by health consumers looking for health information. That type of data has been used to study the information-seeking behaviors of health consumers [30]. Textual comments to YouTube videos have been also found to contain personal health information [31]. There are more examples of textual unstructured data, as we will explain in detail in the two study cases below.

We need to consider also the processing of events from SM sources. For example, the processing of published news about a potential health outbreak can generate another type of data regarding a potential event. In many type of applications it is necessary to process a data streams (e.g., Twitter posts) in order to identify real-time events (e.g., emerging topics). This type of analysis/processing has very unique requirements due to the need of having real-time results. For example, Twitter has been analysed to discover emerging opinions regarding tobacco products [32].

Regarding data acquisition, we need to take into account also the use of sensor data from devices such as step-counters and home sensors. These sensor data sources are increasingly being integrated in SM applications. For example, you can share results from Fitbit with your friends. The location of your device acquired using the GPS sensor is commonly added to any type of SM shared for the mobile phone. Although this type of sensor-based data is not yet very common in the health domain, experts are considering this as one of the areas with more potential in the health domain due to the incoming mass adoption of the Internet of the Things [33].

5.2.2 Data Analysis Processing

It is very common that the data extracted from SM needs to be processed before it can be analyzed. For example, with natural text there are several phases involved. The stop words (extremely common words such as "the") need to be removed since they do not provide valuable information. Then the words needs to be "shortened" to their root in a process defined as stemming. These word stems are used as the source of data for many different types of analysis. For example, word stems can be matched to particular medical ontologies to extract semantic meaning. Semantic analysis has been used to identify medical terms in YouTube videos [34], and also to identify the vocabulary used by patients when searching health information [35].

Another type of analysis is sentiment analysis. In this case the SM content is analyzed to extract cues about the sentiments expressed on it. For example, information about vaccination can include words such as "fear" and "anxiety," which can infer a negative bias toward vaccination [36]. Sentiment analysis has been used to detect the mood of cancer patients in online networks aiming at providing a more personalized feedback [37].

5.2.3 Data Storage and Curation

The data needs to be stored in a secure location for analysis and also developing applications. Nowadays data tend to be stored in systems that are connected to the Internet allowing the interoperability between different cloud-based systems. Often the captured data can be incomplete but it can be inferred from other sources. For example, marketing web forms often do acquire the company of the responder by analyzing the email address. This process is called "data augmentation" and it can be used to acquire additional data in online health interventions [38]. Other pros-processing of the data that can be seen as data curation include data validation (i.e., to identify errors in the data such inconsistencies), data redundancy control (i.e., to avoid duplications in the data).

5.2.4 Data Visualization, Usage, and Services

To provide value to the end users, the data needs to be visualized. That is not a simple task per se since it is unprecedented the visualization of complex massive data sources. Consequently, Data Visualization has become a major area of research within medical informatics [39]. In a recent publication Hesse et al. discussed how social data can be integrated with other data sources for social science research in the health domain [40] (Fig. 5.3).

The visualization of health-related data can go one step further and be transformed into Clinical Decision Support Systems aiming at helping both patients and professionals making decisions. For example, PatientsLikeme.com provides health recommendations to patients based on their information and the information from thousands of patients with similar health problems and treatments. That service, which is described below in a study case, is also used by health researchers to better understand the target population. One of the most complete systems for visualization, simulation, and analytics in

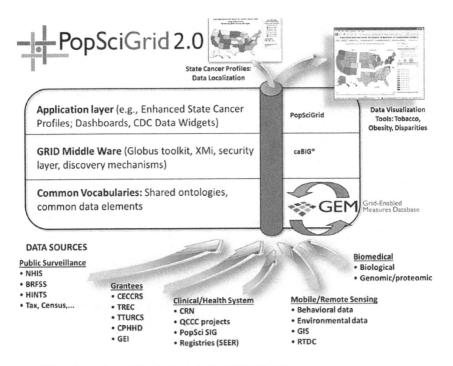

Figure 5.3 Data Integration and Visualization in the PopSciGrid 2.0 [40].

the health domain is the IBM Watson system, which integrates data from many data providers, to offer complex analytics in multiple sectors, including both the health domain [41] and SM [42].

5.2.5 HealthMap: Big Data for Public Health

HealthMap.org is an iconic project and pioneer in the area of Digital Epidemiology [43]. The basic idea behind HealthMap.org is to monitor the massive amount of online news (and SM) for public health surveillance (e.g., health outbreaks, and medical device hazards). The system HealthMap was created by Brownstein in 2007 [44], the same year YouTube was created, and currently it has been expanded to monitor millions of website daily in multiple languages. Fig. 5.4 describes the main elements of the HealthMap system.

The system has been used in multiple settings for the early detection of infectious diseases [46]. It was also used to monitor the H1N1 pandemic which originated in Mexico, Fig. 5.5 shows the alert on the potential outbreak [47].

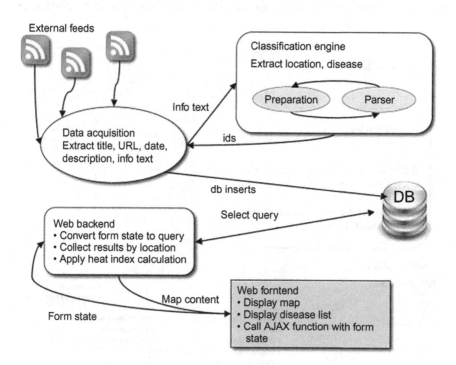

Figure 5.4 HealthMap architecture [45].

Figure 5.5 H1N1 alert in the HealthMap system [47].

Figura 5.6 FluNearYou App [48].

Recently the system is including new data sources using apps designed for patients and citizens, so enriching the data acquired by the web platform. An example is the website www.flunearyou.org that allows user to search for the incidence of influenza in their vicinity after entering detailed information of their symptoms. This approach allows us the collection of detailed health information for the monitoring of the influenza epidemics [48]. Fig. 5.6 shows the interface of the system.

5.3 CHALLENGES OF BIG DATA APPLIED TO SM

To fully take advantage of Big Data applied to SM and health, all the stakeholders involved in the management of data and decision-making, such as health professionals and organizations on the one hand, and patients and general public on the other hand, should collaborate in order to find the best way to make the most of the analysis of the data generated in SM platforms. In general, there are several issues related to Big Data that must be addressed when working out such technologies, in particular technical aspects and ethical considerations that are described below.

5.3.1 Technical Aspects

From the technical point of view, there are several and critical challenges related to Big Data that need to be improved or solved. On the one hand, the information is often heterogeneous and collected in a format that is not ready for analysis and requires specific previous process in order to be manageable. On the other hand, aspects related to the quality, completeness, and reliability of data as well as their reproducibility in similar conditions should be tackled carefully in order to better handle the data and the knowledge derived from it [17]. This point is particularly relevant, above all if the analysis of data is used for decision-making and healthcare management, which is a particularly sensitive issue because of both its nature and the consequences of its potential misinterpretation or misuse. Besides, without the appropriate information technologies infrastructures, requirements, and analytic tools, the potential contributions of Big Data can be misleading or even lost [7,49]. In addition, data integration and representation is an important challenge and the use of more sophisticated visualization techniques and algorithms should be developed to facilitate organizations the interpretation and the practical application of the information obtained. Technical aspects related to security in a Big Data breaches can be big too. In that sense, it is necessary to develop specific strategies to improve the effectiveness of security systems and provide more advanced detection of fraud and illegal activities.

Another important aspect that should be managed is the use, application, and development of standards, vocabularies and ontologies, specifically designed for representing and integrating data across multiple sources, being a key component when dealing with Big Data [50]. In addition, all these changes also make it necessary to set up a training plan for a new generation of scientists with a strong background in statistics, information technologies, and computer science and at the same time with a good knowledge of medicine and biology [51].

5.3.2 Ethical and Legal Considerations When Using SM for Health Purposes

Besides the technical aspects, there are some relevant issues when exploiting health personal Big Data and managing the potential conflicts related to privacy and data protection in general and in particular when managing personal health information that is considered a very sensitive information. Aspects related to the identification of not only the owners of the raw data but also the Big Data processes and outputs are a very important matter that should be faced up. In this sense, for instance, specific regulations and policies to cleanly differentiate public and private information should be established in order to manage the potential risks issues related to Big Data, which should be managed in an appropriate manner. Legal, ethical, and policy issues constitute some very important aspects in the context of sharing human-subjects' protection related to Big Data research techniques. Governments, health organizations, and institutions as well as patients and general public have to agree upon the management of the use of personal health data, setting up clear policies regarding ownership, secondary use, and applications of this data [52]. The use of SM platforms has introduced innovations in healthcare that generate several dilemmas and unanswered legal and ethical questions. In the last few years the nature of the relationship between patients and healthcare professionals has changed with the use of SM as a means of communication and these tools are increasingly used among health students and professionals, sharing health-related content information [53–55]. The incorporation of SM services in the day-to-day healthcare activities should be accompanied by setting up measures to guarantee a safe utilization and thereby ensuring medical professionalism, ethical and legal requirements, particularly on using open and massive online services [56]. To regulate this new scenario, several guidelines and recommendations of good practices has been drawn up in order to point out the most important aspects that should be considered in the use of SM for health purposes, considering the drawbacks and advantages in the use of them [57–61], although there is not yet a general agreement among health professionals on how some situations on these networks should be managed [62,63].

There is a consensus on the potential benefits and opportunities that SM may provide when used for healthcare purposes, which were unthinkable less than a decade ago. The eases of reaching a massive audience in order to promote public health and patient education, to facilitate better and quicker access to healthcare services and professionals, used as a means for patient engagement to promote the communication between patients, relatives, and support groups or to provide low-cost health services and the possibility to recruit participants for the development of clinical trials based on their specific profiles, are some of these benefits [64–66].

Likewise the use of SM has brought unsuspected drawbacks and challenges related to the protection of personal data, particularly in the case of personal sensitive data such as health information. Taking into account that the public display of connections with other users and the content generated is accessible by everyone connected to the Internet, maintaining privacy, confidentiality, the right to be forgotten, and risks related to misinformation are some of the main concerns when posting and participating in these networks. Furthermore, in this scenario the use of sophisticated and automatic tools for collecting and monitoring personal data and behavior has become increasingly common and less detectable. Furthermore, different studies suggest that unprofessional uses of SM are not uncommon and diverse improper behaviors such as posting sexually suggestive photos or criticism of others, diverse violations of patient privacy, are among the most common unprofessional behaviors detected in these platforms [67–70].

To sum up, the use of SM services for health purposes and many tools related to Big Data Analytics are in the process of being widely adopted in healthcare organizations and it is frequent that the legal recommendations or ethical guidelines are behind the introduction of new technical developments and solutions. For this reason, it is essential to promote a wide reflection and that the authorities and governments establish, in collaboration with patients associations and professional institutions, specific ethical, legal guidelines, and use policies to the benefit of the current and future healthcare professional–patient relationship and general public.

5.4 CONCLUSIONS

The possibility of accessing and dealing with health data coming from very different resources such as scientific publications, drug and toxicology databases, all the "omics" data currently available (genomics, metabolomics, and proteomics), EHRs, and in particular, health information that can be extracted from SM platforms and services, is very challenging. By discovering hidden associations and patterns within the data and transforming information into predictive models [2], Big Data analytics has the potential to improve people's health by means of contributing to a more personalized medicine, offering better health services and contributing in the reduction of costs. All these changes can help to strengthen and improve the patient-centered care after decades of disease-centered model of care, paving the way for a customization of healthcare and precision medicine [6]. Nonetheless, initiatives and projects related to Big Data are still at an early stage, and for most of the health institutions and organizations, it is not clear enough what type of data and analysis are of real relevance and value, and what the strategy should be to solve the internal changes needed, with the aim to achieve the adaptations to this new scenario and the continuous process that requires and at the same time in a sustainable way. In conclusion, we are facing emerging scenarios and challenges that offer unsuspected opportunities related to the Big Data Analytics and its application to health care but a close collaboration and interdisciplinary approach is required.

REFERENCES

[1] McAfee A, Brynjolfsson E. Big data: the management revolution. Harv Bus Rev 2012;90 (10):60−128.

[2] Raghupathi W, Raghupathi V. Big Data analytics in healthcare: promise and potential. Health Inf Sci Syst 2014;2(3):1−10.

[3] Costa FF. Big Data in biomedicine. Drug Discov Today 2014;19(4): 443−440.

[4] Li Y, Chen L. Big biological data: challenges and opportunities. Genomics Proteomics Bioinformatics 2014;12:187−9.

[5] Bellazi R. Big Data and biomedical informatics: a challenging opportunity. IMIA Yearbk Med Inform 2014;9:8−13.

[6] Huang T, Lan L, Fang X, An P, Min J, Wang F. Promises and challenges of Big Data computing in health sciences. Big Data Res 2015;2(1):2−11.

[7] Mayer MA. Expectations and pitfalls of Big Data in biomedicine. Rev Pub Admin Manage 2015;3:1.

[8] Big Data Research and Development Initiative. The White House. Office of Science and Technology Policy Executive Office of the President. United States of America. Available at: <https://www.whitehouse.gov/sites/default/files/microsites/ostp/big_data_press_release.pdf>.

[9] European Commission. Big Data strategy. Available at: <http://ec.europa.eu/digital-agenda/en/towards-thriving-data-driven-economy>.

[10] Big Data Value Association. Available at: <http://www.bdva.eu/>.

[11] Greene CS, Tan J, Ung M, Moore JH, Cheng C. Big Data informatics. J Cell Physiol 2014;229:1896–900.

[12] Qin Y, Yalamachili HK, Qin J, Yan B, Wang J. The current status and challenges in computational analysis of genomic Big Data. Big Data Res 2015;2:12–18.

[13] Nash DB. Harnessing the power of Big Data in healthcare. Am Health Drug Benefits 2014;7(2):69–70.

[14] Schneeweiss S. Learning from big health care data. N Engl J Med 2014;370(23):2161–3.

[15] Issa NT, Byers SW, Dakshanamurthy S. Big data: the next frontier for innovation in therapeutics and healthcare. Exp Rev Clin Pharmacol 2014;7(3):293–8.

[16] Mayer MA, Furlong LI, Torre P, Planas I, Cots F, et al. Reuse of EHRs to support clinical research in a hospital of reference. Stud Health Technol Inform 2015;210:224–6.

[17] Cases M, Furlong LI, Albanell J, Altman RB, Bellazi R, et al. Improving data and knowledge management to better integrate health care and research. J Intern Med 2013; 274:321–8.

[18] Signorini A, Segre AM, Polgreen PM. The use of Twitter to track levels of disease activity and public concern in the US during the Influenza A H1N1 pandemic. PLoS One 2011;6(5):e19467.

[19] Kostkova P, de Quincey E, Jawaheer G. The potential of Twitter for early warning and outbreak detection. Available at: <http://science.icmcc.org/2010/04/14/the-potential-of-twitter-for-early-warning-and-outbreak-detection>; 2010.

[20] Chew C, Eysenbach G. Pandemics in the age of Twitter: content analysis of tweets during the 2009 H1N1 outbreak. PLoS ONE 2010;5(11):e14118.

[21] Paul MJ, Dredze M. You are what you tweet: analyzing twitter for public health. In: Adamic LA, Baeza-Yates RA, Counts S, editors. ICWSM. Menlo Park, CA: The AAAI Press; 2011.

[22] Prieto VM, Matos S, Álvarez M, Cacheda F, Oliveira JL. Twitter: a good place to detect health conditions. PLoS One 2014;9(1):e86191.

[23] Carbonell P, Mayer MA, Bravo A. Exploring brand-name drug mentions on Twitter for pharmacovigilance. Stud Health Technol Inform 2015;210:55–9.

[24] Sloane R, Osanlou O, Lewis D, Bollegala D, Maskell S, Pirmohamed M. Social media and pharmacovigilance: a review of the opportunities and challenges. Br J Clin Pharmacol 2015;80(4):910–20.

[25] Paul MJ, Sarker A, Brownstein JS, Nikfarjam A, Scotch M, Smith KL, Gonzalez G. (2016) Social media mining for public health monitoring and surveillance. In: Pacific symposium on biocomputing, Hawaii, USA.

[26] Leis A, Mayer MA, Torres Niño J, Rodriguez-Gonzalez A, Suelves JM, Armayones M. Healthy eating support groups on Facebook: content and features. Gac Sanit 2013; 27(4):355–7.

[27] BDVA. European Big Data Value Strategic Research & Innovation Agenda. Available at: <http://www.bdva.eu/sites/default/files/europeanbigdatavaluepartnership_sria__v1_0_final.pdf>.

[28] Fernandez-Luque L, Karlsen R, Bonander J. Review of extracting information from the Social Web for health personalization. J Med Internet Res 2011;13(1):e15.

[29] Jacobs I (2014) Opensocial Foundation moves standards work to W3C social web activity. Available at: <https://www.w3.org/blog/2014/12/opensocial-foundation-moves-standards-work-to-w3c-social-web-activity/>.

[30] Yom-Tov E, Borsa D, Cox Ingemar J, McKendry RA. Detecting disease outbreaks in mas gatherings using Internet data. J Med Internet Res 2014;16(6):e154.

[31] Fernandez-Luque L, Elahi N, Grajales FJ, III. An analysis of personal medical information disclosed in YouTube videos created by patients with multiple sclerosis. Stud Health Technol Inform 2009;150:292−6.

[32] Myslin M, Zhu SH, Chapman W, Conway M. Using Twitter to examine smoking behavior and perceptions of emerging tobacco products. J Med Internet Res 2013;15(8):e174.

[33] Verzijl D, Dervojeda K, PwC Netherlands and Laurent Prost & Laurent Frideres, PwC Luxembourg. Directorate-General for Internal Market, Industry, Entreneurship and SMEs, Directorate J Industrial Property, Innovation & Standards. European Union (2015) Internet of Things. Smart Health, Business Innovation Observatory. Available at: <http://ec.europa.eu/growth/industry/innovation/business-innovation-observatory/files/case-studies/46-iot-smart-health_en.pdf>.

[34] Konstantinidis S, Fernandez-Luque L, Bamidis P, Karlsen R. The role of taxonomies in social media and the semantic web for health education. A study of SNOMED CT terms in YouTube health video tags. Methods Inf Med 2013;52(2):168−79.

[35] Doing-Harris KM, Zeng-Treitter Q. Computer-assisted of a consumer health vocabulary through mining of social network data. J Med Internet Res 2011;13(2):e37.

[36] Salathé M, Khandelwal S. Assessing vaccination sentiments with online social media: implications for infectious disease dynamics and control. PLoS Comput Biol 2011;7(10):e1002199.

[37] Rodrigues RG, das Dores RM, Camilo-Junior CG, Rosa TC. SentiHealth-Cancer: a sentiment analysis tool to help detecting mood of patients in online social networks. Int J Med Inform 2016;85(1):80−95.

[38] Spek V, Nyklicek I, Smits N, Cuijpers P, Riper H, Keyzer J, et al. Internet-based cognitive behavioural therapy for subthreshold depression in people over 50 years old: a randomized controlled clinical trial. Psychol Med 2007;37(12):1797−806.

[39] Caban JJ, Gotz D. Visual analytics in healthcare—opportunities and research challenges. JAMIA 2015;260−2.

[40] Hesse BW, Moser RP, Riley WT. From Big Data to knowledge in the social sciences. Ann Am Acad Pol Soc Sci 2015;659(1):16−32.

[41] Kohn MS, Sun J, Knoop S, Shabo A, Carmeli B, Sow D, et al. IBM's health analytics and clinical decision support. Yearb Med Inform 2014;9(1):154−62.

[42] IBM Watson Analytics for Social Media. Available at: <https://www.ibm.com/marketplace/cloud/social-media-data-analysis/us/en-us>.

[43] Salathé M, Bengtsson L, Bodnar TJ, Brewer DD, Brownstein JS, Buckee C, et al. Digital epidemiology. PLoS Comput Biol 2012;8(7):e1002616.

[44] Brownstein JS, Freifeld CC. HealthMap: the development of automated real-time internet surveillance for epidemic intelligence. Euro Surveill 29, 2007;12(11):E071129.5.

[45] Freifeld CC, Mandl KD, Reis BY, Brownstein JS. HealthMap: Global infectious disease monitoring through automated classification and visualization of Internet media reports. J Am Inform Assoc 2008;15(2):150−7.

[46] Barboza P, Vaillant L, Le Strat Y, Hartley DM, Nelson NP, Mawudeku A, et al. Factors influencing performance of internet-based biosurveillance systems used in epidemic intelligence for early detection of infectious diseases outbreaks. PLoS One 5, 2014;9(3):e90536.

[47] Brownstein JS, Clark CC, Freifeld BS, Madoff LC. Influenza A (H1N1) virus, 2009—online monitoring. N Engl J Med 2009;360:2156.

[48] FluNearYou App. Available at: <https://itunes.apple.com/es/app/flu-near-you/id570301361?mt=8>.

[49] Roski J, Bo-Linn GW, Andrews TA. Creating Value in health care through Big Data: opportunities and policy implications. Health Aff 2014;33:1115−22.

[50] Merelli I, Pérez-Sánchez H, Gesing S, D'Agostino D. Managing, analysing, and integrating Big Data in medical informatics: open problems and future perspectives. BioMed Res Int 2014;2014: 134023.

[51] Martin-Sánchez F, Verspoor K. Big Data in medicine is driving big changes. IMIA Yearb Med Inform 2014;9:14−20.

[52] Metcalf J. Council for Big Data, Ethics, and Society (2015) Human-subjects protection and Big Data: open questions and changing landscapes. Available at: <http://bdes.datasociety.net/>.

[53] Rodríguez-González A, Mayer MA, Fernández-Breis JT. Biomedical information through the implementation of Social Media environments (Guest Editorial). J Biomed Inform 2013;46:955−6.

[54] Bosslet GT, Torke AM, Hickman SE, Terry CL, Helft PR. The patient-doctor relationship and online social networks: results of a national survey. J Gen Intern Med 2011;26(10):1168−74.

[55] Lewis MA, Dicker AP. Social Media and Oncology: the past, present and future of electronic communication between physician and patient. Semin Oncol 2015;42(5):764−71.

[56] Mayer MA, Leis A, Mayer A, Rodriguez-González A. How medical doctors and students should use social media: a review of the main guidelines for proposing practical recommendations. Stud Health 2012;180:853−7.

[57] World Medical Association Statement on the professional & ethical use of social media (2011). Available from: <http://www.wma.net/en/30publications/10policies/s11/>.

[58] Australian Medical Association Council and the New Zealand Medical Association (2010). Available from: <https://ama.com.au/article/social-media-and-medical-profession>.

[59] Canadian Medical Association (2011) Social media and Canadian physicians—issues and rules of engagement. Available from: <www.cma.ca/socialmedia>.

[60] British Medical Association: Using social media: practical and ethical guidance for doctors and medical students (2013). Available from: <https://bma.org.uk/-/media/files/pdfs/practical advice at work/ethics/socialmediaguidance.pdf>.

[61] American Medical Association Policy in the use of Social Media (2010). Available from: <http://www.ama-assn.org/ama/pub/dab/9124a-abstract.page>.

[62] Thompson LA, Black E, Duff WP, Black NP, Saliba H, Dawson K. Protected health information on social networking sites: ethical and legal considerations. J Med Res 2011;13(1):e8.

[63] Brown AD. Social media: a new frontier in reflective practice. Med Educ 2010;44:744−5.

[64] Denecke K, Bamidis P, Bond C, Gabarron E, Househ M, Lau AYS, et al. Ethical issues of social media usage in healthcare. Yearb Med Inform 2015;10(1):137−47.

[65] Henderson M, Dahynke M. The ethical use of social media in nursing practice. Medsurg Nurs 2015;24(1):62−4.

[66] Denecke K. Ethical aspects of using medical social media in healthcare applications. Stud Health Technol Inform 2014;198:55−62.

[67] MacDonald J, Sohn S, Ellis P. Privacy, professionalism and Facebook: a dilemma for young doctors. Med Educ 2010;44:805−13.

[68] Thompson LA, Dawson K. The intersection of online social networking with medical professionalism. J Gen Intern Med 2008;23(7):954−7.

[69] White J, Kirwan P, Krista L, Walton J, Ross S. 'Have you seen what is on Facebook?' The use of social networking software by healthcare professions students. BMJ Open 2013;3: e003013.

[70] Langenfeld SJ, Cook G, Sudbeck C, Luers T, Schenarts PJ. An assessment of unprofessional behaviour among surgical residents on Facebook: a warning of the dangers of social media. J Surg Educ 2014;71(6):e28−32.

CHAPTER 6

Social Media and Health Behavior Change

L. Laranjo
Macquarie University, Sydney, Australia

6.1 INTRODUCTION

Chronic diseases are nowadays the leading cause of mortality and morbidity worldwide [1–3]. One of the major factors contributing to the global burden of chronic diseases is the "epidemic" of unhealthy lifestyle behaviors [1]. Indeed, sedentariness, poor diet, smoking, and alcohol abuse are common risk factors for cardiovascular diseases and other chronic conditions, accounting for a great amount of healthcare costs.

There is now overwhelming evidence that optimizing lifestyle behaviors is a key factor in both the prevention and management of chronic illnesses. However, the best approach to promote sustained behavior change remains to be identified [4]. Indeed, health behavior change continues to be one of the biggest challenges in modern day life [1,5].

Participatory medicine, a "movement in which networked patients shift from being mere passengers to responsible drivers of their health" [6], is closely connected to the growing importance of health behavior change. Several burdensome chronic conditions nowadays are associated with unhealthy lifestyle behaviors. Therefore, increasing patient participation in care is important to promote the behavioral modifications that are needed for the prevention, management, and treatment of these conditions. Indeed, behavior change is utterly dependent on the individual, hence the importance of promoting patient involvement and engagement with their health and health care.

Patients are increasingly able to use information technology to help them make informed decisions about health care [3,7]. The term "*e-patients*" describes "empowered, engaged, equipped, and enabled" patients, who are able to use modern electronic tools to actively

Participatory Health Through Social Media. DOI: http://dx.doi.org/10.1016/B978-0-12-809269-9.00006-2

participate in care, and to be heard by other patients, physicians, and policy makers [6,8−10].

The number of e-patients is rapidly increasing [10] due to a myriad of factors: the convenience of using the internet to search for health information; the growing popularity of social media and its use for health purposes; and the increased availability of biosensors and self-trackers (e.g., sleep/activity trackers), which led to the development of the quantified self-movement ("the movement of making lifestyle-related decisions based on everyday measurements of health parameters") [6,9,11,12].

6.2 IMPORTANCE OF HEALTH BEHAVIOR CHANGE

Behavior change is relevant in many fields, from primary prevention of chronic diseases to chronic disease management and treatment of mental health problems. Indeed, behavior change interventions have been successfully applied to a wide variety of health behaviors and conditions [13]. Importantly, even small changes from interventions and policies on relevant health behaviors can lead to substantial public health improvements and cost savings in health care [14].

6.2.1 Primary Prevention of Chronic Diseases

Healthy lifestyle behaviors are essential in the prevention of several chronic conditions. It is known that increasing and maintaining physical activity levels is associated with lower mortality from all causes, as well as improvements in quality of life and physical function, and reductions in body weight [15−18]. Also, a balanced diet is important for the prevention of obesity, diabetes, and cardiovascular diseases in general, as well as for the maintenance of a healthy weight [19,20].

6.2.2 Chronic Disease Management

The Chronic Care Model is a well-known approach to improve the quality of care for chronic conditions [3,21]. A central element of the chronic care model is the activated patient, with the knowledge, skills, and confidence to participate in the management of their disease [3,22−25].

Self-management activities involve managing symptoms, doing the necessary treatments, making lifestyle changes, and coping with the physical and psychosocial consequences of the disease, with the aim of minimizing its impact on health.

Research has shown that self-management support and behavior change programs tend to be effective in improving disease knowledge, symptom management, self-management behaviors, self-efficacy, and a variety of clinical outcomes [4,21,26–31].

Elements that are commonly involved in successful self-management and behavioral change programs include: education; collaborative problem definition; self-management training and support; targeting, goal setting, planning, skill development, problem solving; and follow-up [21,28,32–34]. Recently, interest in the use of information and communication technology (ICT) to facilitate self-management and promote patient empowerment has increased considerably, showing favorable results [27,35–45].

6.2.3 Treatment of Mental Health Problems

Cognitive behavioral therapy (CBT) is effective in the management and treatment of a variety of mental health problems, namely anxiety and depression. Interestingly, CBT can be delivered online, with very good results [46,47]. Plus, delivering CBT online makes it widely accessible and convenient, turning it into a cost-effective therapy that is able to benefit a greater number of patients than is possible in traditional CBT.

Online CBT can be self-guided or guided by a clinician, and it may include reminders, access to lessons and educational content [48]. Cognitive behavioral interventions can be delivered online for a variety of mental health problems and behavioral issues, with high adherence and effectiveness [46,47].

6.3 THEORIES, MODELS, AND FRAMEWORKS OF HEALTH BEHAVIOR CHANGE

A theory of behavior change aims to explain why, when, and how a behavior does or does not occur, as well as its influencing factors [14]. Currently, several theories and models can provide insight on the process of behavior change, but no unique behavioral change theory dominates health research and practice [5].

Until date, the most commonly used behavior change theories have been the ones focusing on reflective cognitive processes such as intention, attitudes, and beliefs, namely the Health Belief Model, Social Cognitive Theory, the Theory of Reasoned Action, and the Transtheoretical Model of Change [5,14,49]. However, other important aspects influencing human behavior have received less attention, such as emotions, habits, drives, impulses, self-control, context, and environment, despite their inclusion in several theories and models [14,50].

Given the different constructs covered by each theory or model, it is not unusual to use a combination of theories, models, and frameworks in behavior change interventions [5]. In the end, the application of any of these theories in clinical practice, public health, or research should be guided by the characteristics of the target population and context, and one single intervention may be guided by several different theories in order to cover the specific construct believed to be relevant in that case.

A brief discussion of some health behavior change theories and models will follow, as well possible applications of their constructs in social media interventions.

6.3.1 The Behavior Change Wheel and the COM-B System

A new method for characterizing and designing behavior change interventions is the Behavior Change Wheel, a more comprehensive framework which can be used to identify the psychological, social, and/or environmental domains to target in a behavior change intervention [14,51]. This framework consists of a behavior system involving three essential components: capability, opportunity, and motivation (designated the "COM-B system").

Capability is defined by the authors as "the individual's psychological and physical capacity to engage in the activity concerned" and it involves having the necessary knowledge and skills. **Motivation** is defined as all the brain processes that energize and direct behavior, and it includes both automatic (e.g., habit, emotional responding, impulses) and reflective processes (involving evaluations, plans, goal setting, and analytical decision making). **Opportunity** is defined as all the factors that lie outside the individual that enable or prompt behavior, and may be divided into physical (i.e., environment-related) and social opportunities.

The COM-B system does not place priority in any single component or perspective: individual, group, and environmental perspectives are all considered equally important in influencing behavior, and the system is applicable both to population-level and individual-level interventions. Furthermore, a given intervention might target one or more components in the behavior system, depending on the constructs relevant to address in each specific case.

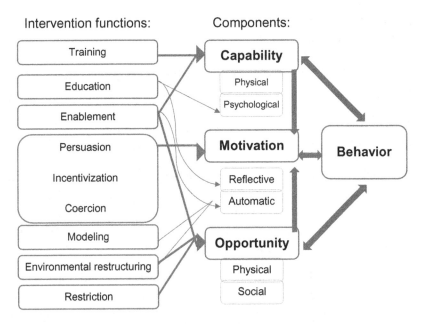

Intervention functions: Components:

Figure 6.1 Schematic representation of the COM-B framework for behavior change. Source: Adapted from Michie S, van Stralen MM, West R. The behavior change wheel: a new method for characterising and designing behavior change interventions. Implement Sci. BioMed Central Ltd 2011;6(1):42.

The COM-B system also includes nine intervention functions (training, education, enablement, persuasion, incentivization, coercion, modeling, environmental structuring, and restriction) which are linked to the system's components (Fig. 6.1). Interestingly, intervention functions can be linked to one or more behavior change techniques (and vice versa), making this framework especially useful as a comprehensive tool in characterizing and evaluating behavior change interventions [51,52].

6.3.2 Health Belief Model

The Health Belief Model proposes that people are most likely to take preventative action if they perceive the threat of a health risk to be serious, if they feel they are personally susceptible and if there are fewer costs than benefits to engaging in it [14]. Therefore, a central aspect of the Health Belief Model is that behavior change interventions are more effective if they address an individual's specific perceptions about susceptibility, benefits, barriers, and self-efficacy [5]. Interventions focusing on this model may involve risk calculation and prediction, as well as personalized advice and education.

6.3.3 Social Cognitive Theory

Social Cognitive Theory (SCT) proposes that the environment, behavior, and personal and cognitive factors all interact as determinants of each other [5,14]. According to this theory, human functioning is described in terms of a number of basic capabilities: symbolizing capability, forethought capability, vicarious capability (ability to learn through observation/imitation/modeling others' behaviors and attitude), self-regulatory capability, and self-reflective capability.

The key concepts of SCT can be grouped into five major categories: (1) psychological determinants of behavior (outcome expectations, self-efficacy, and collective efficacy), (2) observational learning, (3) environmental determinants of behavior (incentive motivation, facilitation), (4) self-regulation, and (5) moral disengagement [5].

An important concept in SCT is **self-efficacy**, which represents a person's belief in their capacity to perform a given behavior when faced with a variety of challenges [53]. According to SCT, self-efficacy may be developed in four ways: (1) personal experience of success, (2) social modeling (showing the person that others like themselves can perform/acquire a certain behavior, as well as the small steps taken by them), (3) improving physical and emotional states, and (4) verbal persuasion (encouragement by others to boost confidence) [5].

Another key concept is **observational learning**, which implies learning to perform new behaviors by exposure to interpersonal or media displays of that same behavior [5]. In this regard, peer modeling is a particularly relevant method for influencing behavior, because imitation occurs more frequently when observers perceive the models as similar to themselves [5].

Finally, according to SCT, **self-regulation** may be achieved in six different ways: (1) self-monitoring and systematic observation of one's own behavior, (2) goal setting, (3) feedback on the quality of performance and how it might be improved, (4) self-reward, (5) self-instruction, and (6) social support from people who encourage a person's efforts to exert self-control [5]. Social support, on the other hand, may be categorized into four types of supportive behaviors [5]: (1) emotional support, involving the provision of empathy, love, trust, and caring; (2) instrumental support, involving the provision of aid, resources, and services that directly assist a person in need; (3) informational support, meaning the provision of advice, suggestions, and information that a person can use to address problems; and (4) appraisal support or the provision of information that is useful for self-evaluation purposes (e.g., constructive feedback) [5].

Therefore, it seems logical that new technologies such as social media are particularly well suited for the application of SCT. Indeed, social media may serve as a source of modeling, verbal persuasion, feedback, and encouragement, as well as may contribute to address other aspects of SCT.

6.3.4 Transtheoretical Model of Behavior Change

The Transtheoretical Model proposes that behavior change occurs in five sequential stages: precontemplation (not planning to change within the next 6 months), contemplation (ambivalent or thinking about change), preparation (taking steps towards changing), action (attempting the change), and maintenance (having been able to sustain behavior change for more than 6 months and working to prevent relapse) [5,14]. Furthermore, 10 "processes of change" are considered important in facilitating movement between different stages, and 2 additional variables influence that process: decisional balance (evaluating the pros and cons of changing) and self-efficacy.

One attraction of stage-based approaches like the Transtheoretical Model, the I-Change Model, the Health Behavior Goal Model and others is that they facilitate tailoring of interventions [14].

6.3.5 I-Change Model (Integrated Change Model)

The I-Change Model integrates several models of behavior change (such as the Theory of Reasoned Action and the Theory of Planned Behavior) and defends that the primary determinant of behavior is a person's intention to carry out that behavior [14]. In this stage-based model, behavior is defined as having two categories ("trial" and "maintenance") and intention as having three different states ("precontemplation," "contemplation," and "preparation"). According to this model, intentions may not always transfer into behavior, depending on ability factors and barriers to action. The primary determinant of intentions is motivation, which in turn is determined by three factors: attitudes, social influences, and self-efficacy. Social influences encompass others' perceptions of the behavior (social norms), observation of others carrying out the behavior (social modeling), and the pressures or support from others to execute the behavior (pressure/support).

As is the case with Social Cognitive Theory, the I-Change Model is particularly well suited for social media interventions, which may contribute to facilitate the effects of social influence.

6.3.6 Goal Setting Theory

Goal Setting Theory explains the mechanisms by which goals influence behavior, and how the latter can be moderated by goal characteristics (difficulty and specificity), the level of commitment, the importance of the goal, levels of self-efficacy, feedback, and task complexity [14].

Interventions focusing on goal setting may help individuals manage each of these aspects to improve chances of success. For instance, a social media intervention may be designed to help increase commitment to a goal by making it public and may facilitate feedback on progress towards a goal, which can in turn improve self-efficacy and performance of the behavior. Additionally, increases in self-efficacy may be promoted through training, role models, and persuasive communication. Finally, task complexity may be reduced by turning a long-term goal into several small goals that may seem more achievable in the short-term, facilitating performance.

6.3.7 Health Behavior Goal Model

The Health Behavior Goal Model is a stage model proposing that behavior change is most likely to occur if the target change is compatible with a person's personal goal structure (what is important to a person and what they want to achieve in life) [14]. According to this model, behavior change is also influenced by the expected consequences of the target behavior, which may be divided into four categories: perceived health costs and benefits, perceived emotional costs and benefits, social influence (expectations of how the social environment might respond to the adoption of the target behavior), and perceived competence. On the other hand, a person's personal goal structure can be altered by environmental characteristics and personal characteristics, which can act as cues for behavior change.

6.3.8 Integrated Theory of Health Behavior Change

The Integrated Theory of Health Behavior Change aims to explain the adoption of self-management behaviors. Motivation is seen as a necessary precursor of behavior change and three main factors are proposed to influence behavior change: knowledge and beliefs, self-regulatory skills and abilities, and social facilitation.

Self-regulation is defined as the process by which people incorporate behavior change into their everyday lives, and it involves: self-monitoring, goal setting, reflective thinking, decision making, planning, plan enactment, self-evaluation and management of emotions arising as a result of behavior change. Furthermore, the concept of **social facilitation** incorporates social influence (influence of a credible source on a person's thoughts and motivation to change) and social support (emotional, instrumental, and informational).

According to this theory, knowledge alone is believed to be insufficient to lead to health behavior change. Nevertheless, knowledge and beliefs can influence engagement in self-regulatory activities, which in turn can lead to improvements in self-management behavior. Furthermore, social facilitation may positively influence both self-regulation and self-management behavior [14].

Social media interventions may be designed to reflect several constructs of the Integrated Theory of Health Behavior Change. For example, sharing behavioral data with others in a network may facilitate self-regulation by outsourcing the job of monitoring, easing the burden on people's daily lives, and enabling people to benefit from feedback and encouragement from the network [54]. Also, social media interventions are particularly suited to promote social influence and social support in behavior change efforts.

6.3.9 Health Promotion Model

The Health Promotion Model aims to explain the factors underlying motivation to engage in health-promoting behaviors and it focuses on people's interactions with their physical and interpersonal environments during attempts to improve health [14]. This model emphasizes the active role that a person has in initiating and maintaining health-promoting behavior, and in shaping their own environment to support health-promoting behaviors.

Factors influencing health-promoting behavior are divided into three categories: "individual characteristics and experiences," "behavior-specific cognitions and affect," and "behavioral outcome."

The model also describes eight behavior-specific beliefs which are believed to determine the health-promoting behavior and are proposed as targets for behavior change interventions: (1) perceived benefits of action, (2) perceived barriers to action, (3) perceived self-efficacy, (4) activity-related affect, (5) interpersonal influences (including norms, modeling/vicarious learning, and social support), (6) situational influences, (7) commitment to plan of action, and (8) immediate competing demands and preferences (alternative behaviors that compete for a person's attention and time) [14].

Social media interventions may target one or more of these behavior-specific beliefs, being particularly useful in addressing interpersonal influences and commitment to a plan of action.

6.3.10 PRIME Theory

The PRIME Theory of Motivation provides a framework in which more specific theories of choice, self-control, habits, emotions, and drives are combined [14]. According to this theory, the human motivational system is constituted by five subsystems: response co-ordination, impulses/inhibition, motives (wants and needs), evaluations (beliefs about what is good or bad), and plans (self-conscious intentions). These subsystems interact with each other and are influenced by the internal and external environments.

Furthermore, the processes of change are divided into automatic (not requiring self-conscious thought) or reflective (inference and analysis). Automatic processes include perception, memory, habituation, sensitization (becoming more responsive with repeated occurrences of a stimulus), associative learning (e.g., operant and classical conditioning), imitation, dissonance reduction (forming or changing beliefs to reduce emotional or motivational conflict), maturation, and physical and chemical interventions.

Finally, PRIME Theory views identity as a very important factor and as the source of self-control (i.e., acting in accordance with self-conscious plans when facing competing desires, impulses, and inhibitions arising from other sources).

Social media interventions may be designed to address the subsystems of the human motivational system as well as facilitate the process of imitation or promote self-control. Self-control can be contagious because it is influenced by social proof—people can be influenced to have more willpower if they believe that self-control is the norm in their social environment [55]. This may be used in an intervention as a strategy to reduce the self-control demands of behaving in a certain desired manner.

6.3.11 Pressure System Model
The Pressure System Model is a theory of behavior change that aims to provide a guide for behavior change counseling in **primary care**, proposing that behavior change is determined by two opposed sources of pressure: motivation and resistance to change [14]. Being based upon the Transtheoretical Model, it classifies people into five stages of the behavior change process, suggesting counseling strategies for each.

Motivation is considered to have three components: beliefs about importance (how important the health condition to be avoided is), beliefs about personal risk (perceptions of one's personal risk of experiencing the health condition), and beliefs about the efficacy of change (perceptions of whether change will lead to the desired health outcome). Sources of resistance to behavior change include both internal and external obstacles: capability to change, locus of control and fixed impediments to change.

Social media interventions may be designed in a tailored manner, addressing the motivational components, the sources of resistance to behavior change, or both.

6.3.12 Extended Information Processing Model
The Extended Information Processing Model aims to provide an explanation of the processes underlying attitude and behavior change resulting from **mass media campaigns** [14]. Five communication aspects are considered important: factors relating to the source of the message, the message content and style, the channel used to transmit the message, the message audience, and the issues being targeted. Additionally, 12 steps are described in the process of behavior change: (1) Exposure to the message, (2) Awareness, (3) Knowledge, (4) Memory (of the message content), (5) Changes in beliefs, (6) Retention of new beliefs, (7) Changes in attitude, (8) Persistence (retaining the attitude change),

(9) Intentions (to engage in the wanted behavior), (10) Resistance (of intentions to change), (11) Behavior (acting on the basis of attitude/intention to change), and (12) Maintenance of behavior changes.

Social media seems to be a potentially good venue for mass media campaigns, allowing for cost-efficiency, easy implementation, scalability, and wide dissemination.

6.3.13 Extended Parallel Processing Model

The Extended Parallel Processing Model aims to explain the cognitive processes and behaviors that occur in response to **fear appeals**, as well as to identify the factors that determine an adaptive or maladaptive response to a perceived threat [14].

Fear appeal messages have four possible components: susceptibility (the likelihood of being vulnerable to the threat), severity (the seriousness of the threat), self-efficacy (one's ability to take recommended protective action), and response efficacy (the likelihood that taking protective action will be effective) [14].

If both perceived efficacy and perceived threat are high, people are motivated to devise strategies to avoid the threat (adaptive changes), defined as a "danger control process." On the contrary, if perceived threat is high but perceived efficacy low, a "fear control process" will be initiated: the initial fear is magnified and people try to cope with it by using fear-reducing strategies (maladaptive changes) such as denial, dissonance, or rejection of the message. Appraisal of threat and efficacy is influenced by a person's prior experiences, personality characteristics, and culture.

Interventions involving fear appeals may be disseminated through social media. However, unless specific strategies to promote self-efficacy are devised, there is a considerable risk of promoting maladaptive changes instead of the intended change in behavior.

6.3.14 Social Change Theory and Social Networks

Social Change Theory proposes that the external environment influences community goals, norms (shared rules and expectations), and values, which influence behavioral change at an individual level [14]. The theory assumes that health improvements are best achieved by altering norms at the community level, rather than at the individual level. According to Social Change Theory, social norms that lead to unhealthy behaviors can be replaced with social norms that support healthier behaviors. Consequently, as changes in the social environment bring about new norms, widespread individual behavior change becomes more likely to occur.

In this theory, change is spread to groups within the community through a process of diffusion via social networks, by which people are exposed to changing norms and through which reinforcement may occur via influential role models.

A **social network** is the web of social relationships that surround individuals [56]. Social networks enable several social functions: social influence, social control, social comparison, companionship, and social support [5]. Recently, **social networking sites** (SNSs) adapted the concept of "social networks" to the online world [57]. This phenomenon will be discussed later in the chapter.

In social networks, tie strength is a measure of the importance of the interactions between two individuals. Strong ties are a major source of social support but weaker ties are also important, as they seem to be better at providing new information and new contacts which may influence behavior (a phenomenon known as the "strength of weak ties," described by Granovetter) [56].

Obesity, smoking, alcohol consumption, depression, and other health behaviors and conditions seem to "spread" through social networks [58–61]. Indeed, many observational studies have shown an association between those conditions and the composition of an individual's social network. For instance, weight gain and obesity seem to be associated with social aspects and behaviors within a social network [62,63]. It has been estimated that the rate of becoming obese increases by 0.5 percentage points for each obese social contact an individual has [58], suggesting that the norms and behaviors that lead to obesity may indeed propagate through social contagion [48]. Not only that, but eating habits also appear to be socially contagious [48,64].

There are three possible explanations for the clustering of weight patterns in social networks: social contagion, social selection, and demographic confounders. Social contagion may occur due to the existence of social norms, which define the behaviors that are more acceptable, according to how frequently they are displayed by elements in the social network [48]. Social selection occurs when people associate with others who share similar interests, beliefs, and behaviors, a tendency that is known as homophily. On the other hand, demographic confounders such as culture, ethnicity, and socioeconomic status characterize the environment where individuals reside and develop their social networks, leading to exposure to similar events, which can often explain apparent associations between network membership and individual characteristics [48].

Interestingly, **network interventions** try to enhance existing networks, develop new networks, identify natural helpers, build community capacity, or combine these strategies to modify social networks.

Furthermore, interventions involving social media (e.g., social networking sites) may promote social comparison by allowing individuals to compare their behaviors with others, as well as may facilitate social influence and diffusion of behaviors through peers and other contacts in a social network [54].

6.3.15 Other Useful Concepts in Behavior Change

Recently, frameworks for behavior change are moving away from pushing people to change into reducing obstacles to change, **nudging**, and promoting informed decision making [5]. Nudging, a concept born from the field of behavioral economics, means attempting to alter peoples' behavior in directions that will make their lives better, by changing particular aspects of the choice architecture in a given situation, therefore making it easier for them to make good decisions [55,65].

Social media interventions may be used to nudge people in directions consistent with their values and goals and to "prime" individuals to increase the ease with which certain information comes to mind [65]. Additionally, the power of social influence can be used strategically to nudge people to perform a behavior that conforms with what "most people are already doing," by using perceived social norms to influence behavior in the direction of the norm [65].

Finally, since focusing on long-term benefits is not a natural human tendency, several strategies have been proposed to overcome willpower failures—turning behaviors into habits, improving choice design through nudging, using rewards, and outsourcing willpower (e.g., having a running buddy or personal trainer) are just a few of the strategies that might be used [66]. **Habits** are behaviors that become automatic when a choice made at some point is repeated consistently without thinking [67], and habit formation may be promoted through social media interventions that reinforce and facilitate the repetition of healthy behaviors.

6.4 BEHAVIORAL INFORMATICS INTERVENTIONS

6.4.1 Definition

Behavioral informatics interventions assist individuals (patients or healthy consumers) in modifying behaviors to improve physical, mental, or behavioral health [48]. These interventions have the potential to be highly cost-effective therapeutic options in the management of several diseases and health problems. Nowadays, the growth in social media brings new opportunities to increase accessibility and use of behavior change interventions.

Behavioral informatics interventions have been applied in several health domains, from wellness and lifestyle improvement (e.g., healthy diet, fitness, smoking cessation), disease management (e.g., asthma, type 2 diabetes, hypertension), and treatment of mental health conditions (e.g., anxiety disorders) [48,68–70]. Tailored feedback and interactive monitoring are common features of behavioral informatics interventions [48,68–70].

A Cochrane review found that computer-based interactive health communication applications (IHCAs) are able to improve cognitive and social support outcomes in patients with chronic conditions [36]. Furthermore, interactive online interventions have also been associated with increases in physical activity, nutritional knowledge, slower health decline, and improved body shape perception [71].

6.4.2 Social Media

6.4.2.1 Importance in Behavioral Medicine

Social media may be defined as "a group of internet-based applications that build on the ideological and technological foundations of Web 2.0, and that allow the creation and exchange of user-generated content" [10]. The use of social media for health purposes is gaining interest, namely the use of blogs, discussion boards, wikis and, especially, SNSs [72–74].

Social media may be particularly suited for behavioral informatics interventions, helping shape social processes associated with behavior change [48]. Social media facilitates access to a community of peers, which may help in providing support, information, and access to resources to help in self-care and self-management of a chronic condition [48]. Indeed, it is known that peer support groups can influence decisions and actions [75–77] and that social media interventions can facilitate disease management by creating online spaces where patients are able to interact with clinicians and between themselves [48].

On the other hand, social media have the potential to be an effective channel to broadcast public health messages and to engage citizens in health promotion and disease prevention behaviors [78].

Interestingly, the social networks created in online communities can be as diverse as other human structures, varying between loose, tight, open, close, opportunistic, or secretive [48]. Indeed, groups of individuals who have short-term common purposes can come together in an online social network and share knowledge, support, resources, and experience, influencing each other.

SNSs can be defined as web-based services that allow individuals to create a personal profile and build a list of connections to other users, originating innumerous interconnected and dynamic personal networks [57].

SNSs are becoming ubiquitous in people's everyday life, making them especially appealing in the public health domain. On one hand, they present a low-cost opportunity to virally spread health information, possibly improving the cost-effectiveness of health interventions. On the other hand, they can promote social support and social influence, facilitating health behavior change.

Finally, the exponential uptake of SNSs offers a new approach to chronic care [12]. As an example, the site PatientsLikeMe helps patients with chronic conditions better understand their disease and therapeutic options, while offering connections to patient communities [79].

6.4.2.2 Use of Social Networking Sites in Promoting Health Behavior Change

SNSs [57] are now a global phenomenon. As of September 2013, 73% of online adults were using an SNS of some kind and 42% were using more than one [80,81]. Facebook is the most popular platform (with more than 1.19 billion monthly active users [82]), followed by Twitter (500 million users worldwide [83]). In parallel to general-purpose SNSs like Facebook and Twitter, health-specific SNSs are also emerging [84,85]. Some are oriented towards patients with a specific chronic condition (e.g., TuDiabetes [86,87]), others are more general and designed for patients with any chronic condition (e.g., PatientsLikeMe [88,89]), and a few others target people wanting to change a particular health risk behavior or improve their lifestyle (e.g., smoking cessation [90]).

The application of SNSs in the health domain shows tremendous potential both at the population and individual levels [91]. At the population level, they are currently being used for public health surveillance [92] for communicable [92,93] and noncommunicable diseases [86,87]. At the individual level, they are able to facilitate access to health-related information [94−97] and social support [56,90], promoting better-informed treatment decisions [88,89].

Given that lifestyle behaviors are nowadays responsible for the global burden of noncommunicable diseases [1], growing attention is focusing on how to use SNSs to counteract this trend [60,98]. Interestingly, studies of offline social networks have demonstrated the actual role of social influence in spreading certain risk behaviors such as in the case of alcohol consumption [99], smoking [59], and obesity [58].

Research efforts are now focusing on how to leverage social influence to promote healthy behaviors. The fact that SNSs are widely accessible across geographical barriers, and that they are increasingly being used by people on a daily basis (namely through mobile phones), turn them into especially interesting loci for public health interventions in the behavioral domain.

Recently, a published meta-analysis evaluating the effectiveness of SNSs in changing health behavior outcomes found a statistically significant positive effect of SNS interventions on behavior change, boosting encouragement for future research in this area (Fig. 6.2) [100].

The results of this meta-analysis are in line with what has been shown for Interactive Health Communication Applications (IHCAs), where a positive effect on behavioral outcomes was found in a Cochrane meta-analysis [36]. IHCAs are comparable to SNS interventions in that the former are computer-based (usually web-based) systems combining health information with either social support, decision support, or behavior change support, and the latter tend to

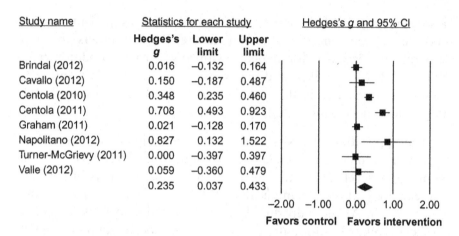

Figure 6.2 *Forest plot of effect sizes and 95% confidence intervals representing the effect of interventions with social networking sites on health behavior-related outcomes (random effects model).* Source: Reproduced with permission from Laranjo L, Arguel A, Neves AL, Gallagher AM, Kaplan R, Mortimer N, et al. The influence of social networking sites on health behavior change: a systematic review and meta-analysis. J Am Med Inform Assoc 2015;22(1):243–56.

provide education, social support, and self-management support [100]. The combination of these functions has also been previously described as being commonly used in other social media interventions [72].

Finally, other systematic reviews have been published evaluating the effect of social media (e.g., blogs, discussion boards, wikis) on health behavior change [72], health promotion [73], and health communication [74], showing feasibility but no definitive conclusion regarding effectiveness.

6.4.3 SNS Intervention Characteristics

Until date, fitness (e.g., physical activity and weight loss promotion) has been the predominant health domain addressed in studies of SNSs for behavior change [100], which reflects the growing interest of consumer informatics research in this field [101,102]. In the future, as more patients with chronic illnesses become social media users [95,103], it is expected that SNS research will increasingly focus on chronic disease self-management.

Until now, the type of SNS being used in behavior change interventions has predominantly been a general SNS like Facebook [105,106,110–114] and Twitter [107], but some studies published to date have also used health-specific SNSs [108,109,115,116]. General SNSs present several advantages for the implementation of health interventions, compared to health-specific SNSs [117,118]. They have enormous reach—millions of regular users worldwide [80,82]—potentially minimizing problems of retention and lack of adherence to interventions. Also, they can be efficient ways of disseminating interventions and recruiting participants [106,111,113,114], and they can take advantage of participants' existing social networks [119,120], instead of asking them to form new connections (which has been termed "the stranger phenomenon" [101]). Finally, as general SNSs are nowadays a part of people's daily lives, they have a huge potential to improve engagement, by being easily incorporated in people's routines and habits, instead of being an extra burden [121]. Indeed, retention rates of general SNS studies are very promising, shedding new light on the "law of attrition" of online interventions [72,100,122].

The majority of interventions involving SNSs for health behavior change seem to consist of other components in addition to the SNS, most often in the form of a website [100]. The scarcity of single component interventions has been previously reported regarding social media and other web-based interventions [101–104], posing problems in determining the effectiveness of a particular component. It is unclear whether the observed effects in studies with multi-component interventions are attributable to either the SNS or the non-SNS component, or to a synergistic effect of both. Furthermore, until date, the majority of studies have not reported on the effects of individual features of an intervention on effectiveness, engagement, or user satisfaction [100].

Another aspect regarding SNS interventions is that the use of a specific health behavior theory or model underlying the intervention has not been common practice [105–109]. There is now considerable evidence showing that interventions grounded in theory can lead to more powerful effects [5,68] but few authors seem to take these theories and models into consideration when designing interventions. A possible consequence of this is that resources may continually be wasted in nonoptimized and "nonevidence-based" interventions.

6.4.4 Network Interventions
Two innovative studies involving the use of an SNS for behavior change have also been published [108,109] using "network alteration" [123].

In those studies, the interventions were based on two aspects of offline social networks: the tendency of people to associate with ones who resemble them—homophily [124]; and the tendency for people's friends to be connected between them, through redundant ties—clustering [124,125]. The author hypothesized that people were more likely to adopt a behavior if they knew that someone similar to them, or some of their friends' friends, had done it before [126]. By modifying participants' networks in an SNS it was indeed demonstrated that homophily and clustering contribute to the social diffusion of "easy" behaviors (e.g., adoption of a diet diary). Nonetheless, it remains to be demonstrated that the same mechanism applies to more complex health behaviors (e.g., dieting, exercising, smoking cessation) [127]. Indeed, it is known that the need for social reinforcement increases when the adoption of a given behavior is difficult, costly, or unfamiliar [126].

6.4.5 Future Avenues

Interventions for health behavior change involving general and health-specific SNSs are feasible and show promise. Future research should focus on identifying the features that increase engagement and retention of the target audience, as well as the specific characteristics that promote long-term behavior change and improve cost-effectiveness.

One aspect that is gaining increasing attention is the possibility of incorporating social features, particularly social networking, in PHRs [128,129], as these may boost adoption [130] and contribute to improvements in health behaviors [100,131]. The Australian PHR "Healthy.me" already incorporates social features, which were found by users to be the most useful and engaging features in this PHR [132].

An interesting hypothesis—that remains untested—is that SNSs may be used in a synergistic way with Personal Health Records and mobile devices [128,129], allowing consumers to continuously benefit from the daily knowledge, accountability, support, and influence that their social connections can provide.

6.5 RISKS ASSOCIATED WITH THE USE OF SOCIAL MEDIA FOR HEALTH BEHAVIOR CHANGE

Despite all the advantages and benefits of SNSs, it is important to consider potential risks and unintended consequences of the use of social media in health behavior change interventions. One of the concerns regards the issue of quality and validity of the information disseminated in social media. A study of diabetes social networks has found a considerable variation in scientific validity of the information disseminated, as well as in the moderation and auditing of those discussions [133].

Another risk is that social media may be used for marketing purposes (e.g., tobacco, alcohol, direct-to-consumer advertisement of medications) or may influence people through public displays of unhealthy behavior (e.g., pro-anorexia, self-injury, drug use) [134].

Finally, social media may have a negative psychological impact on individuals, when used to propagate offensive, inappropriate, or stigmatizing content.

6.6 CONCLUSION

In this chapter, several theories for behavior change were discussed, as well as their possible application in social media interventions. In particular, the use of SNSs for health behavior change was analyzed in more detail, and the characteristics of these interventions were explored. Specific challenges in the implementation of social media interventions for behavior change were also mentioned, as well as the potential risks that social media can bring about in the realm of health behaviors and behavior change.

REFERENCES

[1] Narayan KMV, Ali MK, Koplan JP. Global noncommunicable diseases—where worlds meet. N Engl J Med 2010;363(13):1196–11988.

[2] Daar AS, Singer PA, Persad DL, Pramming SK, Matthews DR, Beaglehole R, et al. Grand challenges in chronic non-communicable diseases. Nature 2007;450(November):494–6.

[3] Institute of Medicine. Crossing the Quality Chasm: a new health system for the 21st century. Building a Better Delivery System; 2001.

[4] Norris S, Engelgau M, Narayan K. Effectiveness of self-management training in type 2 diabetes: a systematic review of randomized controlled trials. Diabetes Care 2001;24(3):561–87.

[5] Glanz K, Rimer B, Viswanath K. Health behavior and health education. 4th ed. San Francisco: John Wiley & Sons; 2008.

[6] DeBronkart D. Let patients help! CreateSpace Independent Publishing Platform; 2013.

[7] Rigby M, Ronchi E, Graham S. Evidence for building a smarter health and wellness future—key messages and collected visions from a Joint OECD and NSF workshop. Int J Med Inform 2013;82(4):209–19 Elsevier Ireland Ltd. Available from: http://dx.doi.org/10.1016/j.ijmedinf.2012.10.003.

[8] deBronkart D. From patient centred to people powered: autonomy on the rise. BMJ 2015;350(February 10 14):h148.

[9] Meskó B. The guide to the future of medicine: technology and the human touch. Dr. Bertalan Meskó; 2014. p. 274.

[10] Meskó B. Social media in clinical practice. London: Springer; 2013. p. 155.

[11] Topol EJ. The patient will see you now: the future of medicine is in your hands. Philadelphia: Basic Books; 2015. p. 384.

[12] Topol E. The creative destruction of Medicine. New York: Basic Books; 2013. p. 336.

[13] Michie S. Designing and implementing behaviour change interventions to improve population health. Health Serv Res 2008;13(October):64–70.

[14] Michie S, West R, Campbell R, Brown J, Gainforth H. ABC of behaviour change theories; 2014.

[15] Gregg EW, Cauley JA, Stone K, Thompson TJ, Bauer DC, Cummings SR, et al. Relationship of changes in physical activity and mortality among older women. JAMA [Internet] 2003;289 (18):2379–86. Available from: <http://www.ncbi.nlm.nih.gov/pubmed/12746361>.

[16] Montgomery P, Dennis J. Physical exercise for sleep problems in adults aged 60+. Cochrane Database Syst Rev 2002;4:CD003404. Available from: <http://www.ncbi.nlm.nih.gov/pubmed/12519595>.

[17] Irwin ML, Yasui Y, Ulrich CM, Bowen D, Rudolph RE, Schwartz RS, et al. Effect of exercise on total and intra-abdominal body fat in postmenopausal women: a randomized controlled trial. JAMA 2003;289(3):323–30. Available from: <http://www.ncbi.nlm.nih.gov/pubmed/12525233>.

[18] Fransen M, Mcconnell S. Exercise for osteoarthritis of the knee (review). Cochrane Libr 2015;1:1–94.

[19] Lindström J, Ilanne-Parikka P, Peltonen M, Aunola S, Eriksson JG, Hemiö K, et al. Sustained reduction in the incidence of type 2 diabetes by lifestyle intervention: follow-up of the Finnish Diabetes Prevention Study. Lancet 2006;368(9548):1673–9. Available from: <http://www.ncbi.nlm.nih.gov/pubmed/17098085>.

[20] Estruch R, Ros E, Salas-Salvadó J, Covas M-I, Corella D, Arós F, et al. Primary prevention of cardiovascular disease with a Mediterranean diet. N Engl J Med 2013;368(14):1279–90. Available from: <http://www.nejm.org/doi/abs/10.1056/NEJMoa1200303>.

[21] Wagner EH, Austin BT, Von Korff M. Organizing care for patients with chronic illness. Milbank Q 1996;74:511–44.

[22] Hibbard JH, Mahoney ER, Stock R, Tusler M. Do increases in patient activation result in improved self-management behaviors?. Health Serv Res 2007;42(4):1443–63.

[23] Hibbard JH, Tusler M. Assessing activation stage and employing a "next steps" approach to supporting patient self-management. J Ambul Care Manage 2007;30(1):2–8.

[24] Bodenheimer T. Improving primary care for patients with chronic illness. JAMA 2002; 288(14):1775.

[25] Coleman K, Austin BT, Brach C, Wagner EH. Evidence on the Chronic Care Model in the new millennium. Health Aff 2009;28(1):75–85.

[26] Hibbard JH. Moving toward a more patient-centered health care delivery system. Health Aff (Millwood) 2004;(Suppl. Var):VAR133–5.

[27] Solomon M, Wagner SL, Goes J. Effects of a Web-based intervention for adults with chronic conditions on patient activation: online randomized controlled trial. J Med Internet Res 2012;14(1):e32.

[28] Newman S, Steed L, Mulligan K. Self-management interventions for chronic illness. Lancet 2004;364(3):1523–37.

[29] Lorig KR, Sobel DS, Stewart AL, Brown BW, Bandura A, Ritter P, et al. Evidence suggesting that a chronic disease self-management program can improve health status while reducing hospitalization: a randomized trial. Med Care 1999;37(1):5–14.

[30] Bodenheimer T, Lorig K, Holman H, Grumbach K. Patient self-management of chronic disease in primary care. JAMA 2002;288(19):2469–75.

[31] Tricco A, Ivers N, Grimshaw J, Moher D, Turner L, Galipeau J. Effectiveness of quality improvement strategies on the management of diabetes: a systematic review and meta-analysis. Lancet 2012;379(9833):2252–62.

[32] Lorig KR, Holman H. Self-management education: history, definition, outcomes, and mechanisms. Ann Behav Med 2003;1–7.

[33] Hill-Briggs F. Problem solving in diabetes self-management: a model of chronic illness self-management behavior. Ann Behav Med 2003;25(3):182–93.

[34] Coleman MT, Newton KS. Supporting self-management in patients with chronic illness. Am Fam Physician 2005;72(8):1503–10.

[35] Samoocha D, Bruinvels DJ, Elbers NA, Anema JR, van der Beek AJ. Effectiveness of web-based interventions on patient empowerment: a systematic review and meta-analysis. J Med Internet Res 2010;12(2):1–21.

[36] Murray E, Burns J, See TS, Lai R, Nazareth I. Interactive health communication applications for people with chronic disease. Cochrane Database Syst Rev 2005;(4):CD004274 Chichester, UK: John Wiley & Sons, Ltd; Sep 1 [cited 2014 May 27]. Available from: <http://www.ncbi.nlm.nih.gov/pubmed/16235356>.

[37] Bull SS, Gaglio B, McKay HG, Glasgow RE. Harnessing the potential of the internet to promote chronic illness self-management: diabetes as an example of how well we are doing. Chronic Illn 2005;1:143–55.

[38] Nagykaldi Z, Aspy CB, Chou A, Mold JW. Impact of a wellness portal on the delivery of patient-centered preventive care. J Am Board Fam Med 2012;25(2):158–67.

[39] Jackson CL, Bolen S, Brancati FL, Batts-Turner ML, Gary TL. A systematic review of interactive computer-assisted technology in diabetes care: Interactive information technology in diabetes care. J Gen Intern Med 2006;21:105–10.

[40] Cho JH, Chang SA, Kwon HS, Choi YH, Ko SH, Moon SD, et al. Long-term effect of the internet-based glucose monitoring system on HbA1c reduction and glucose stability: a 30-month follow-up study for diabetes management with a ubiquitous medical care system. Diabetes Care 2006;29(12):2625–31.

[41] Costa BM, Fitzgerald KJ, Jones KM, Dunning Am T. Effectiveness of IT-based diabetes management interventions: a review of the literature. BMC Fam Pract 2009;10:72.

[42] Lorig K, Ritter PL, Laurent DD, Plant K, Green M, Jernigan VBB, et al. Online diabetes self-management program. Diabetes Care 2010;33(6):1275–81.

[43] Cotter AP, Durant N, Agne AA, Cherrington AL. Internet interventions to support lifestyle modification for diabetes management: a systematic review of the evidence. J Diabetes Complications 2014;28:243–51.

[44] McMahon GT, Gomes HE, Hohne SH, Hu TMJ, Levine BA, Conlin PR. Web-based care management in patients with poorly controlled diabetes. Diabetes Care 2005;28:1624–9.

[45] Ralston JD, Revere D, Robins LS, Goldberg HI. Patients' experience with a diabetes support programme based on an interactive electronic medical record: qualitative study. BMJ 2004;328(7449):1159.

[46] Proudfoot J, Klein B, Barak A, Carlbring P, Cuijpers P, Lange A, et al. Establishing guidelines for executing and reporting internet intervention research. Cogn Behav Ther 2011;40(2):82–97.

[47] Andrews G, Cuijpers P, Craske MG, McEvoy P, Titov N. Computer therapy for the anxiety and depressive disorders is effective, acceptable and practical health care: a meta-analysis. PLoS One 2010;5(10):e13196.

[48] Coiera E. Guide to health informatics. 3rd ed. CRC Press; 2015. p. 710.

[49] Prestwich A, Sniehotta FF, Whittington C, Dombrowski SU, Rogers L, Michie S. Does theory influence the effectiveness of health behavior interventions? Meta-analysis. Health Psychol 2014;33(5):465–74. Available from: <http://www.ncbi.nlm.nih.gov/pubmed/23730717>.

[50] Glanz K, Bishop DB. The role of behavioral science theory in development and implementation of public health interventions. Annu Rev Public Health 2010;31:399–418.

[51] Michie S, van Stralen MM, West R. The behaviour change wheel: a new method for characterising and designing behaviour change interventions. Implement Sci 2011;6(1):42 BioMed Central Ltd

[52] Michie S, Fixsen D, Grimshaw JM, Eccles MP. Specifying and reporting complex behaviour change interventions: the need for a scientific method. Implement Sci 2009;4:40. Available from: <http://www.pubmedcentral.nih.gov/articlerender.fcgi?artid=2717906&tool=pmcentrez&rendertype=abstract>.

[53] Bandura A. Social foundations of thought and action: a social cognitive theory. Englewood Cliffs, NJ: Prentice Hall; 1986.

[54] Baumeister RF, Tierney J. Willpower: why self-control is the secret to success. Penguin Books; 2012.

[55] McGonigal K. The willpower instinct. USA: Avery Publishing Group Inc.; 2013.

[56] Valente T. Social networks and health. Oxford University Press; 2010.

[57] Boyd DM, Ellison NB. Social network sites: definition, history, and scholarship. J Comput Commun 2007;13(1):210–30.

[58] Christakis NA, Fowler JH. The spread of obesity in a large social network over 32 years. N Engl J Med 2007;357(4):370–9.

[59] Christakis NA, Fowler JH. The collective dynamics of smoking in a large social network. N Engl J Med 2008;358(21):2249–58. Available from: <http://www.pubmedcentral.nih.gov/articlerender.fcgi?artid=2822344&tool=pmcentrez&rendertype=abstract>.

[60] Fowler JH, Christakis NA. Dynamic spread of happiness in a large social network: longitudinal analysis over 20 years in the Framingham Heart Study. BMJ 2008;337(337):a2338.

[61] Christakis NA, Fowler JH. Social contagion theory: examining dynamic social networks and human behavior. Stat Med 2013;32(4):556–77. Available from: <http://doi.wiley.com/10.1002/sim.5408>.

[62] Cunningham SA, Vaquera E, Maturo CC, Venkat Narayan KM. Is there evidence that friends influence body weight? A systematic review of empirical research. Soc Sci Med 2012;75(7):1175–83 Elsevier Ltd. Available from: <http://linkinghub.elsevier.com/retrieve/pii/S0277953612004522>.

[63] Fletcher A, Bonell C, Sorhaindo A. You are what your friends eat: systematic review of social network analyses of young people's eating behaviours and bodyweight. J Epidemiol Community Heal [Internet] 2011;65(6):548–55. Available from: <http://jech.bmj.com/cgi/doi/10.1136/jech.2010.113936>.

[64] Pachucki MA, Jacques PF, Christakis NA. Social network concordance in food choice among spouses, friends, and siblings. Am J Public Health 2011;101(11):2170−7.

[65] Thaler RH. Nudge: improving decisions about health, wealth, and happiness. Penguin Books; 2009. p. 312.

[66] Ariely D. Predictabily irrational. In.

[67] Duhigg C. The power of habit; 2014.

[68] Webb TL, Joseph J, Yardley L, Michie S. Using the internet to promote health behavior change: a systematic review and meta-analysis of the impact of theoretical basis, use of behavior change techniques, and mode of delivery on efficacy. J Med Internet Res 2010;12(1)e4 January [cited 2014 May 27]. Available from: <http://www.pubmedcentral.nih. gov/articlerender.fcgi?artid=2836773&tool=pmcentrez&rendertype=abstract>.

[69] van den Berg MH, Schoones JW, Vliet Vlieland TPM. Internet-based physical activity interventions: a systematic review of the literature. J Med Internet Res 2007;9(3):e26.

[70] Stevens VJ, Funk KL, Brantley PJ, Erlinger TP, Myers VH, Champagne CM, et al. Design and implementation of an interactive website to support long-term maintenance of weight loss. J Med Internet Res 2008;10(1):13−26. Available from: <http://ezproxy.staffs.ac.uk/>.

[71] Wantland DJ, Portillo CJ, Holzemer WL, Slaughter R, McGhee EM. The effectiveness of web-based vs. non-web-based interventions: a meta-analysis of behavioral change outcomes. J Med Internet Res 2004;6(4).

[72] Williams G, Hamm MP, Shulhan J, Vandermeer B, Hartling L. Social media interventions for diet and exercise behaviours: a systematic review and meta-analysis of randomised controlled trials. BMJ Open 2014;4(2):e003926.

[73] Chou WS, Prestin A, Lyons C, Wen K. Web 2.0 for health promotion: reviewing the current evidence. Am J Public Health 2013;103(1):e9−18.

[74] Moorhead SA, Hazlett DE, Harrison L, Carroll JK, Irwin A, Hoving C. A new dimension of health care: systematic review of the uses, benefits, and limitations of social media for health communication. J Med Internet Res 2013;15(4):e85.

[75] Lieberman MA, Golant M, Giese-Davis J, Winzlenberg A, Benjamin H, Humphreys K, et al. Electronic support groups for breast carcinoma: a clinical trial of effectiveness. Cancer 2003;97(4):920−5.

[76] Lorig KR, Laurent DD, Deyo RA, Marnell ME, Minor MA, Ritter PL. Can a Back Pain E-mail Discussion Group improve health status and lower health care costs?: a randomized study. Arch Intern Med 2002;162(7):792−6.

[77] Shaw BR, McTavish F, Hawkins R, Gustafson DH, Pingree S. Experiences of women with breast cancer: exchanging social support over the CHESS computer network. J Health Commun 2000;5(2):135−59.

[78] Neiger BL, Thackeray R, Van Wagenen SA, Hanson CL, West JH, Barnes MD, et al. Use of social media in health promotion: purposes, key performance indicators, and evaluation metrics. Health Promot Pract 2012;13(2):159−64.

[79] Braunstein ML. Health informatics in the cloud [Internet]. New York, NY: Springer; 2013. Available from: <http://link.springer.com/10.1007/978-1-4614-5629-2>.

[80] Pew Research Center. Social media update. Available from: <http://www.pewinternet.org/ 2013/12/30/social-media-update-2013/>; 2013.

[81] Pew Research Center. Social networking fact sheet. Available from: <http://www.pewinternet. org/fact-sheets/social-networking-fact-sheet/>; 2013.

[82] Facebook Newsroom. Available from: <http://newsroom.fb.com/content/default.aspx? newsareaid=22>; 2014.

[83] Twitter, by the numbers. Available from: <http://news.yahoo.com/twitter-statistics-by-the-numbers-153151584.html>; 2013.

[84] Korda H, Itani Z. Harnessing social media for health promotion and behavior change. Health Promot Pract 2013;14(1):15−23.

[85] Eggleston EM, Weitzman ER. Innovative uses of electronic health records and social media for public health surveillance. Curr Diab Rep 2014;14.

[86] Mandl KD, McNabb M, Marks N, Weitzman ER, Kelemen S, Eggleston EM, et al. Participatory surveillance of diabetes device safety: a social media-based complement to traditional FDA reporting. J Am Med Inform Assoc 2014;21(4):687−91.

[87] Weitzman ER, Kelemen S, Quinn M, Eggleston EM, Mandl KD. Participatory surveillance of hypoglycemia and harms in an online social network. JAMA Intern Med 2013;173(5):345−51.

[88] Wicks P, Massagli M, Frost J, Brownstein C, Okun S, Vaughan T, et al. Sharing health data for better outcomes on PatientsLikeMe. J Med Internet Res 2010;12(2):e19.

[89] Wicks P, Vaughan TE, Massagli MP, Heywood J. Accelerated clinical discovery using self-reported patient data collected online and a patient-matching algorithm. Nat Biotechnol 2011;29(5):411−14 Nature Publishing Group.

[90] Cobb NK, Graham AL, Abrams DB. Social network structure of a large online community for smoking cessation. Am J Public Health 2010;100(7):1282−9.

[91] Coiera E. Social networks, social media, and social diseases. BMJ 2013;346(May 22 16):f3007.

[92] Eysenbach G. Infodemiology and infoveillance: framework for an emerging set of public health informatics methods to analyze search, communication and publication behavior on the Internet. J Med Internet Res 2009;11(1):e11.

[93] Salathé M, Freifeld CC, Mekaru SR, Tomasulo AF, Brownstein JS. Influenza A (H7N9) and the importance of digital epidemiology. N Engl J Med 2013;369(5):401−4.

[94] Hawn C. Take two aspirin and tweet me in the morning: how Twitter, Facebook, and other social media are reshaping health care. Health Aff (Millwood) 2009;28(2):361−8.

[95] Greene JA, Choudhry NK, Kilabuk E, Shrank WH. Online social networking by patients with diabetes: a qualitative evaluation of communication with Facebook. J Gen Intern Med 2011;26(3):287−92.

[96] Greaves F, Ramirez-Cano D, Millett C, Darzi A, Donaldson L. Harnessing the cloud of patient experience: using social media to detect poor quality healthcare. BMJ Qual Saf 2013;22(3):251−5.

[97] Rozenblum R, Bates DW. Patient-centred healthcare, social media and the internet: the perfect storm? BMJ Qual Saf 2013;22(3):183−6.

[98] Smith KP, Christakis NA. Social networks and health. Annu Rev Sociol 2008;34(1):405−29.

[99] Rosenquist J, Murabito J, Fowler J, Christakis N. The spread of alcohol consumption behavior in a large social network. Ann Intern Med 2010;152:426−33.

[100] Laranjo L, Arguel A, Neves AL, Gallagher AM, Kaplan R, Mortimer N, et al. The influence of social networking sites on health behavior change: a systematic review and meta-analysis. J Am Med Inform Assoc 2015;22(1):243−56.

[101] Chang T, Chopra V, Zhang C, Woolford SJ. The role of social media in online weight management: systematic review. J Med Internet Res 2013;15(11):e262 Jan [cited 2014 May 27]. Available from: <http://www.pubmedcentral.nih.gov/articlerender.fcgi?artid=3868982&tool= pmcentrez&rendertype=abstract>.

[102] Neve M, Morgan PJ, Jones PR, Collins CE. Effectiveness of web-based interventions in achieving weight loss and weight loss maintenance in overweight and obese adults: a systematic review with meta-analysis. Obes Rev 2010;11(4):306–21 Apr [cited 2014 May 27]. Available from: <http://www.ncbi.nlm.nih.gov/pubmed/19754633>.

[103] Hamm MP, Chisholm A, Shulhan J, Milne A, Scott SD, Given LM, et al. Social media use among patients and caregivers: a scoping review. BMJ Open 2013;3(5) 1–10 January [cited 2014 May 27]. Available from: <http://www.pubmedcentral.nih.gov/articlerender. fcgi?artid=3651969&tool=pmcentrez&rendertype=abstract>.

[104] Eysenbach G, Powell J, Englesakis M, Rizo C, Stern A. Health related virtual communities and electronic support groups: systematic review of the effects of online peer to peer inter-actions. BMJ 2004;328(4):1166.

[105] Mayer A, Harrison J. Safe eats: an evaluation of the use of social media for food safety education. J Food Prot 2012;75(8):1453–63 August [cited 2014 May 27]; Available from: <http://www.ncbi.nlm.nih.gov/pubmed/22856569>.

[106] Valle CG, Tate DF, Mayer DK, Allicock M, Cai J. A randomized trial of a Facebook-based physical activity intervention for young adult cancer survivors. J Cancer Surviv 2013;7(3):355–68 September [cited 2014 May 27]. Available from: <http://www.ncbi.nlm. nih.gov/pubmed/23532799>.

[107] Turner-McGrievy G, Tate D. Tweets, apps, and pods: results of the 6-month Mobile Pounds Off Digitally (Mobile POD) randomized weight-loss intervention among adults. J Med Internet Res 2011;13(4):e120 January [cited 2014 May 27]. Available from: <http://www.pubmedcen-tral.nih.gov/articlerender.fcgi?artid=3278106&tool=pmcentrez&rendertype=abstract>.

[108] Centola D. The spread of behavior in an online social network experiment. Science 2010;329(5996):1194–7 September 3 [cited 2014 May 23]. Available from: <http://www. ncbi.nlm.nih.gov/pubmed/20813952>.

[109] Centola D. An experimental study of homophily in the adoption of health behavior. Science 2011;334(6060):1269–72 December 2 [cited 2014 May 27]. Available from: <http:// www.ncbi.nlm.nih.gov/pubmed/22144624>.

[110] Foster D, Linehan C, Kirman B, Lawson S, Gary J. Motivating physical activity at work: using persuasive social media for competitive step counting. ACM 2010;(6):111–16 [cited 2014 May 27]. Available from: <http://dl.acm.org/citation.cfm?id=1930510>

[111] Napolitano MA, Hayes S, Bennett GG, Ives AK, Foster GD. Using Facebook and text messaging to deliver a weight loss program to college students. Obesity 2012;21(1):25–31 April 24 [cited 2014 May 27]. Available from: <http://doi.wiley.com/10.1038/oby.2012.107>.

[112] Cavallo DN, Tate DF, Ries AV, Brown JD, Devellis RF, Ammerman AS. A social media-based physical activity intervention. Am J Prev Med 2012;43(5):527–32.

[113] Bull S, Levine D, Black S, Schmiege S, Santelli J. Social media-delivered sexual health intervention—a cluster randomized trial. Am J Prev Med 2012;43(5):467–74.

[114] Young SD, Jaganath D. Online social networking for HIV education and prevention: a mixed-methods analysis. Sex Transm Dis 2013;40(2):162–7 February [cited 2014 May 24]. Available from: <http://www.pubmedcentral.nih.gov/articlerender.fcgi?artid= 3869787&tool=pmcentrez&renertype=abstract>.

[115] Brindal E, Freyne J, Saunders I, Berkovsky S, Smith G, Noakes M. Features predicting weight loss in overweight or obese participants in a web-based intervention: randomized trial. J Med Internet Res 2012;14(6):e173 January [cited 2014 May 27]. Available from: <http://www.pubmedcentral.nih.gov/articlerender.fcgi?artid=3558051&tool= pmcentrez&rendertype=abstract>.

[116] Graham A, Cobb N, Papandonatos G, Moreno J, Kang H, Tinkelman D. A randomized trial of internet and telephone treatment for smoking cessation. Arch Intern Med

2011;171(8):46–53 [cited 2014 May 27];. Available from: <http://archpsyc.jamanetwork.com/article.aspx?articleid=226380>.

[117] Cobb NK, Graham AL. Health behavior interventions in the age of facebook. Am J Prev Med 2012;43(5):571–2. Available from: <http://www.ncbi.nlm.nih.gov/pubmed/23079184>.

[118] Bennett GG, Glasgow RE. The delivery of public health interventions via the Internet: actualizing their potential. Annu Rev Public Health 2009;30:273–92 January [cited 2014 May 27]. Available from: <http://www.ncbi.nlm.nih.gov/pubmed/19296777>.

[119] Poirier J, Cobb NK. Social influence as a driver of engagement in a web-based health intervention. J Med Internet Res 2012;14(1):e36 January [cited 2014 May 27]. Available from: <http://www.pubmedcentral.nih.gov/articlerender.fcgi?artid=3374540&tool=pmcentrez&rendertype=abstract>.

[120] Rice E. The positive role of social networks and social networking technology in the condom-using behaviors of homeless young people. Public Health Rep 2010;125(4):588–95 [cited 2014 May 27]. Available from: <http://www.ncbi.nlm.nih.gov/pmc/articles/PMC2882610/>.

[121] Jimison H, Gorman P, Woods S, Nygren P, Walker M, Norris S, et al. Barriers and drivers of health information technology use for the elderly, chronically ill, and underserved. Evid Rep Technol Assess (Full Rep) 2008;(175):1–1422. Available from: <http://www.ncbi.nlm.nih.gov/pubmed/19408968>.

[122] Eysenbach G. The law of attrition. J Med Internet Res 2005;7(1):e11 January [cited 2014 May 27]. Available from: <http://www.pubmedcentral.nih.gov/articlerender.fcgi?artid=1550631&tool=pmcentrez&rendertype=abstract>.

[123] Valente TW. Network interventions. Science 2012;337(6090):49–53 July 6 [cited 2014 May 27]. Available from: <http://www.ncbi.nlm.nih.gov/pubmed/22767921>.

[124] McPherson M, Smith-Lovin L, Cook J. Birds of a feather: homophily in social networks. Annu Rev Sociol 2001;27:415–44.

[125] Centola D, Macy M. Complex contagions and the weakness of long ties. Am J Sociol 2007;113(3):702–34 [cited 2014 May 27]. Available from: <http://www.jstor.org/stable/10.1086/521848>.

[126] Centola D. Social media and the science of health behavior. Circulation 2013;127(21):2135–44. Available from: <http://www.ncbi.nlm.nih.gov/pubmed/23716382>.

[127] van der Leij MJ. Sociology. Experimenting with buddies. Science 2011;334(6060):1220–1 December 2 [cited 2014 May 27]. Available from: <http://www.ncbi.nlm.nih.gov/pubmed/22144610>.

[128] Eysenbach G. Medicine 2.0: social networking, collaboration, participation, apomediation, and openness. J Med Internet Res 2008;10(3):e22.

[129] Paton C, Hansen M, Fernandez-Luque L, Lau AY. Self-tracking, social media and personal health records for patient empowered self-care. Contribution of the IMIA Social Media Working Group. Yearb Med Inform 2012;7(1):16–24. Available from: <http://www.ncbi.nlm.nih.gov/pubmed/22890336>.

[130] Lau AY, Dunn AG, Mortimer N, Gallagher A, Proudfoot J, Andrews A, et al. Social and self-reflective use of a Web-based personally controlled health management system. J Med Internet Res 2013;15(9):e211 January [cited 2014 May 4]. Available from: <http://www.pubmedcentral.nih.gov/articlerender.fcgi?artid=3785989&tool=pmcentrez&rendertype=abstract>.

[131] Maher CA, Lewis LK, Ferrar K, Marshall S, De Bourdeaudhuij I, Vandelanotte C. Are health behavior change interventions that use online social networks effective? A systematic review. J Med Internet Res 2014;16(2):e40 January [cited 2014 May 27]. Available from: <http://www.pubmedcentral.nih.gov/articlerender.fcgi?artid=3936265&tool=pmcentrez&rendertype=abstract>.

[132] AY L, Dunn A, Mortimer N, Proudfoot J, Andrews A, ST L, et al. Consumers' online social network topologies and health behaviours. Stud Heal Technol Inf 2013;192:77–81. Available from: <http://search.bvsalud.org/portal/resource/en/mdl-23920519>.

[133] Weitzman ER, Cole E, Kaci L, Mandl KD. Social but safe? Quality and safety of diabetes-related online social networks. J Am Med Informatics Assoc 2011;18(3):292–7. Available from: <http://jamia.oxfordjournals.org/cgi/doi/10.1136/jamia.2010.009712>.

[134] Lau A, Sintchenko V, Crimmins J, Magrabi F, Gallego B, Coiera E. Impact of a web-based personally controlled health management system on influenza vaccination and health services utilization rates: a randomized controlled trial. J Am Med Informatics Assoc 2012;19:719–27.

CHAPTER 7

Gamification and Behavioral Change: Techniques for Health Social Media

P.D. Bamidis[1], E. Gabarron[2,3], S. Hors-Fraile[4,5], E. Konstantinidis[1], S. Konstantinidis[6], and O. Rivera[7]

[1]Aristotle University of Thessaloniki, Thessaloniki, Greece [2]University Hospital of North Norway, Tromsø, Norway [3]The Arctic University of Norway, Tromsø, Norway [4]University of Seville, Spain [5]Salumedia Tecnologías, Seville, Spain [6]The University of Nottingham, Nottingham, UK [7]University of Seville, Seville, Spain

7.1 INTRODUCTION

In this chapter, we will discuss about games, game techniques, and its implications for health and health social media. It will be summarized first the existing evidence on how games have been related to health (pros and cons); then it will be presented the different uses of game techniques to improve healthcare outcomes, and the techniques used for health social media and its effects on behavioral change.

Before reviewing the different game techniques, let's have a look at the background of this new field, and let's think What is a game? The Oxford Dictionary defines game as *"an activity that one engages in for amusement."* In recent years, with the emerge of the new technologies, most of the games have become digital or computerized, they are also known as online games or videogames. The digital games are then, by definition, a form of play using an interactive computer-based game software. These games are easily accessible through different technological platforms such as laptops, smartphones, and tablet computers.

These digital games have been related to other benefits and also to disadvantages in connection with different arenas than the amusement, and specifically are associated with healthcare. And new concepts such as *gamification, serious games*, or *social games* have emerged (Table 7.1).

Before we delve into the different gamification techniques, its impact for behavioral change, and the techniques for health social media, in Section 1 we present a general overview of the literature on

Participatory Health Through Social Media. DOI: http://dx.doi.org/10.1016/B978-0-12-809269-9.00007-4

Table 7.1 Concepts and Definition	
Concepts	Definition
Game	Activity that one engages in for amusement
Digital game	Form of play using an interactive computer-based game software
Gamification	Application of typical elements of game playing, such as points, scores, or competition with others; to other areas of activity different from amusement (i.e., marketing, health education)
Serious game	Form of play using interactive computer-based game software, specifically designed for a purpose other than pure entertainment. Serious games aim to train or educate people, engaging them in their own healthcare
Social games	Refers to playing online games that allow or require social interaction between players, as opposed to playing games in solitude
Social network games	Type of online game that is played through social networks and typically features multiplayer and asynchronous gameplay mechanics

how digital games have been related to health. We introduce the potential of games applied to health. We can find research studies proving how they can harm our health, and also how they can improve it.

7.1.1 Games and Their Potential to Harm

Most of the scientific publications on digital games and health have focused on their potential for harming our health. And in this sense, there are some researches that have demonstrated the link between playing games and its effect on mental health. For example, the exposure to violent games has been related to aggressive cognitions and aggressive behaviors, and also to desensitization to violence, and to a less social behavior [1]. And playing in group, such as multiplayer games, when these games are of violent content has been related to higher rates of aggressive and delinquent behavior [2]. Racing games are reported to increase risk-taking inclinations, sensations seeking, and attitudes toward reckless driving [3]. Other reported unwanted effects of playing games on mental health are the game addiction and also depressive symptoms that have been associated with the time spent playing games.

Besides its effects on mental health, researchers also have found that games can have an impact on general health. And, as example, playing games has been associated to inactivity and development of obesity due to a higher sedentary behavior because of the time spent playing [4]. And the excess of time dedicated to playing games also has been related to poor school performance [5]. More described unwanted effects that have been linked to the use of digital games are the appearance of seizures, motion sickness, and even musculoskeletal pain syndromes related to an excess of computer and video game use [6].

7.1.2 Games and Their Benefits for Health

Although the use of games and gamification for health is a field that is still in its infancy, and more research is needed, there are a new series of publications with promising results in which it is shown that these tools can be beneficial for health. And these benefits are mostly found in relation with the use of serious games or games specifically developed for a purpose other than pure entertainment. The reported advantages can be grouped on games for health education or health promotion targeting the general public; and games for patient self-management, aiming to improve health outcomes in people with a medical condition. In both cases, the use of the gamification strategy help people to learn more about health and how to take care of their conditions while playing.

7.1.2.1 For Health Education or Health Promotion

Serious games have been used as a way to promote health. One of the uses of the gamification on health education has been in relation to the sexual health. This strategy offer an important advantage for sexual health promotion: the anonymity and confidentiality, which are crucial for very sensitive topics as the sexual health [7–10]. And the use of serious games and gamification strategy it seems to be a powerful tool to engage young populations, who have the highest rates of sexually transmitted infections worldwide according to the World Health Organization. Interventions using serious games for sexual health promotion have proved to have significant positive effects in improving knowledge, however, due to the novelty of this speciality, still there are not enough studies to demonstrate its effect over a sexual behavior change [7].

An example on how the gamification is being used for sexual health promotion can be found in the web app: www.sjekkdeg.no [9,10]. This is a site with a game esthetic that is accessible through laptop, smart-phone, and tablet computer. Because privacy is extremely important when concerning sensitive topics such as sexually transmitted diseases, one of the used gamification techniques comprises the use of avatars. These avatars are a cartoon or game-like figure that can be customized by the user; and these avatars act in representation of the user during navigation in the virtual space and can interact with permanent inhabitants like the doctor, the teacher, and a wizard. The site counts with three different scenarios: a clinic, a school, and a cinema. The clinic, which includes interactive game-like functions to teach on sexual health and sexually transmitted diseases; and a symptom checker where the virtual doctor asks questions about symptoms that the avatar has noticed and finally suggests a diagnosis, and also recommends that the avatar should go to the doctor in real life and get tested. The school also has interactive game-like functionality where the avatar can interact with the virtual teacher. The content on the school consists of lessons (educational text and photographs on anatomy, physiology, contraception, and behavioral issues) and also has quizzes that are linked to the lessons. The third scenario of the gamified site is a cinema, which comprises short instructional videos covering topics such as how to put on a condom, coming out, or what happens in a typical appointment at the venereology unit.

Additional gamification techniques used in this game are achievement-based gifts that unlock complements to customize avatar by fulfilling different tasks (e.g., watch videos, read FAQ sections) or challenges between users by sharing quiz results on the social media [9] (Fig. 7.1).

Gamification strategies for health promotion have also been used in the area of mental health. Specifically, these strategies were successfully used to increase the knowledge in young people [11]. Serious

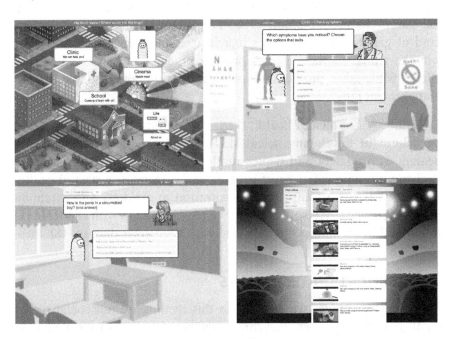

Figure 7.1 Case example of a gamified web app for sexual health promotion.

games for well-being and healthy lifestyle promotion also have proved to have positive effects on their users, specially on improving their knowledge [12], and researchers have proposed to use this method for tobacco prevention and control campaigns [13].

7.1.2.2 For Patient Self-Management

Another use of the games that has proved to have a benefit on health is to educate or improve self-management skills on people that already have been diagnosed or suffer a specific medical condition.

In the area of obesity or weight management, research studies have found that games can be powerful tools in terms of motivation for exercise, which might have an effect on weight reduction [14]. An playing partner exergames also has found to have an effect on obesity prevention, as the fact of playing with a partner make gamers do not perceive their exertion to be higher [15].

Various research studies involving serious games to improve self-management skills in people with diabetes have proved its benefits [16]. These games typically involve players in problem solving and decision making in simulations of diabetes self-management, typically users are asked to balance food and insulin in order to keep the game character's blood glucose within normal range [16]. This kind of games requires players to rehearse skills repeatedly until they win the game, so these games provide practice and show cause and effect, while also providing basic information about diabetes self-management [16].

In the area of mental health, the use of serious games has proved to be helpful in developing social skills in people with autism and in dealing with anxiety and depression [6]. Current research projects are studying more potential benefits of the use of games on mental health, such as its power to improve concentration in children with attention deficit/hyperactivity disorder [17] or potential to help with the rehabilitation and training in people affected with Alzheimer disease [18].

7.2 GAMIFICATION TECHNIQUES

The game techniques we describe in this section are not social media on their own. However, they are strongly related to social media. In fact, game environments are considered as a social media site by

Ventola [19]. Gamification is also present in other social media types such as social networks and content production channels, as described by Lee Ventola. However, we will abstract them in the gamification concept. Sebastian Deterding defined it as the use of game design elements in nongame contexts [20]. As a key component of social media, we will identify how we can better match it to perform controlled behavioral changes.

Actually, gamification and behavioral change are two terms that are linked. After Deterding, Gartner redefined the gamification concept as "the use of game mechanics and experience design to digitally engage and motivate people to achieve their goals" [21]. In other words, it is a practice to encourage people to perform a behavior change. Afterwards, Huotari and Hamari defined gamification as "a process of enhancing services with motivational affordances in order to invoke gameful experiences and further behavioral outcomes" [22,23]. This definition emphasizes the relationship between gamification and user behavior. Currently, a new definition of gamification has been proposed by Robson et al.: "the application of game design principles in order to change behaviors in nongame situations" [24]. Robson et al. identified the behavioral change as the main objective of gamification. Several studies prove that gamified systems increase users' engagement, which matches the proposed definition. However, none use psychological constructs or base the gamification design on existing behavioral change techniques.

The next points will help us understand a bit more the influence gamification and game techniques have on social media, and how they relate with behavioral change techniques. Although they aim the same objective of "behavioral change," it is frequently overlooked the existing synergy between behavioral change techniques and gamification.

7.2.1 Gamification and Game Mechanics

Gamification gurus such as Gabe Zichermann, Andrzej Marczewski, and Yu-Kai Chou have proposed different gamification frameworks. These help us gamify systems in a more structured way, with better understanding of what impact the game components will be on the users, and avoiding what Margaret Robertson called "pointsification"—giving points for interacting with your system with no real game design behind it. Actually, the people misconception that gamification is just points, badges, and leaderboards, triad is still present, and the previously mentioned gamification experts strive to break it down.

Miller et al. [25] made a generic classification of game mechanics in mHealth apps: badges, leaderboards, points and levels, challenges and quests, social engagement loops, and onboarding. However, there are more and all of them can be used in social media. As Sergio Jiménez reflects on his Game On! Toolkit there are more game mechanics such as avatars (player virtual representations), countdowns (they urge players to perform actions before the time runs out), virtual rewards, catalogue, and virtual economy, inventories, progress bars, and random elements (which involve surprises and unexpected results).

Anyway, the game mechanics classification differs from author to author. For example, Yu-Kai Chou expands the possible mechanics much more in his Octalysis framework (Fig. 7.2).

The context of the system which is going to be gamified is a key for success. The range of contexts in which gamification has been implemented

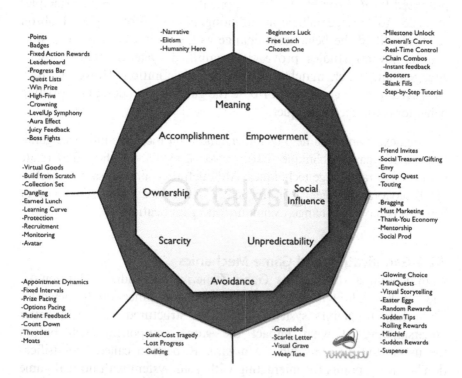

Figure 7.2 Octalysis framework by Yu-Kai Chou. Source: www.yukaichou.com, Image used under authorization.

is wide, including education, commerce, health, intraorganizational system, academic dissemination, government services, public engagement, etc. However, gamification may not be effective in all contexts. As Hamari published [26], it may fail rising interest in the gamified features of the system. He also suggested "service oriented towards strictly rational behavior, such as e-commerce sites, might prove to be challenging systems to be gamified as the users could be geared towards optimizing economic exchanges." In this chapter, we will focus on the health context.

Several studies in the area of gamification applied to exercise and health can be found in the literature. The results of these studies indicate positive effects of gamification, e.g., on physical activity, healthy eating habits, and willingness to continue using the health-related system. For instance, the study performed by Yu Chen et al. "Healthy Together" [27] is a mobile fitness app to motivate people to exercise by setting challenges to groups of two. Players were put in a competitive, collaborative, or hybrid setting to reach exercise goals using gamification such as progress bars, points, poke, badges, bragging, and group quests. The outcomes of this study show that exercise performance depended on the setting in which players were collaborative and hybrid settings had better results than competitive settings (21.1% and 18.2% vs 8.8%, respectively). Another example is the study of Allam et al. [28] The website ONESELF for Rheumatoid Arthritis patients is combined both online social support and gamification mechanics (points, fixed intervals, rewards, progress bars, leaderboards, and badges), which had a positive impact on the empowerment of patients. These outcomes contrast with Camerini and Schulz's on ONESELF, which did not included gamification. This differentiating factor may have positively influenced the Allam et al. patients' behavior and empowerment results.

Therefore, although gamification can be a catalyzer for social media benefits in health, we need to specially plan gamification designs in social environments, and not rely on just adding some mechanics to the system we want to make more engaging. We ask ourselves, is it possible that those cases in which gamification was not successful were poor game design solutions? That the user target and aims of the systems were misaligned with the game mechanics that were developed?

People who use the system are another essential factor for gamification. Studies point towards a heterogeneous set of reasons and behaviors of users who interact with a gamified system. Therefore that variety hinders the assessment of the impact on triggering our motivations. Although more research is needed about this, as far as we know there is no scientific evidence concluding that some people cannot benefit from gamification, it needs to be designed to trigger the motivation of those people. In order to engage the maximum amount of people when their motivations or "player type" is known, it is recommended to implement a range of gamification techniques oriented towards all different player types, so that none is left behind.

Before we start the design of any gamified system, we have to know whether the ingredients we include are suitable for the taste of our target public. There are different models which classify users in different player types. The most popular one is also the simplest one. It was proposed by Richard Bartle. It describes people as multidimensional entities. Based on how we enjoy acting and interacting with the world and other users, we can be "killers," "achievers," "socializers," and "explorers" in different degrees at the same time. It is worth mentioning that the social dimension is the most common present in people, which is especially beneficial to exploit if we are in a social media environment. However, this classification is more oriented to gamers and may lack of the diversity of other models like the Hexad [29].

Amy Jo Kim discussed how to create compelling experiences using gamification in a speech she gave in 2011. She proposed four types of social engagement adapting Bartle's gamer type, and related some action verbs to them.

- Express, which corresponds to the "Achievers" player type: Choose, customize, layout, design, dress-up
- Compete, which corresponds to the "Killers" player type: Win, beat, brag, taunt, challenge, pass, fight
- Explore, which corresponds to the "Explorers" player type: View, read, search, collect, complete, curate
- Cooperate, which corresponds to the "Socializers" player type: Join, share, help, gift, greet.

These verbs, and the proposed groups they are classified in, are of extreme utility to focus a gamified design. Once we know the dynamics we want our system have (the verbs), we can apply the previously mentioned mechanics. With them, we will be able to guide our users through the game experience journey.

7.2.2 Behavioral Change and Gamification

As we mentioned above, the recent definitions of gamification emphasize the existing relationship between user behaviors and gamification. Gamification aims to engage users to change their behaviors in order to acquire and adherence to, e.g., a healthy behavior. Behavioral outcomes have been examined in many empirical studies on gamification as Hamari et al. [30] reported. In general, it seems that it exists a relationship between the affordances of the system and behavioral change. However, there is a lack of evidence of this relationship and further studies measuring latent psychological variables should be conducted to attain more accurate linkages between game mechanics, psychological effects, and behavioral outcomes.

Some authors have suggested these behavioral changes could be caused for the feeling of novelty and curiosity towards the gamification mechanics. Assuming this hypothesis, when the novelty wears off, the changed behavior levels also decrease, and gamification stops being effective. Therefore, gamification has only short-term benefits without long-term effects. However, this fact has not been directly measured in related studies and therefore future research on the effects of novelty in gamification should be performed.

There exist different behavior change theories and taxonomies. Some of them are: the Health Belief Model (HBM), the Social Cognitive Theory (SCT), the Theory of Reasoned Action (TRA), the Theory of Planned Behavior (TPB), and Common Sense Model (CSM). All of these theories have been consolidated into the Appeal Belonging Commitment (ABC) Framework. Other theories that are used in health field are: Self-Determination Theory (SDT), the Transtheoretical Model/Stages of Change (TTM), and social support and social networks. Both individual and social factors are included in these validated health behavior theories.

Gamification aims to enhance services with motivational affordances to invoke experiences and behavioral outcomes. User motivation is the key factor to create desired behavioral change. SDT is a theory that identifies two types of motivations, intrinsic and extrinsic. Intrinsic motivation is related to user's inner values and extrinsic motivation is related to external rewards such as money or social status. Specifically, intrinsic motivation is the kind of motivation in which the activity is rewarding in and of itself, i.e., intrinsic motivation is related to those enjoyable activities that people wish or like to perform on their own sake. Behavioral changes can be created by both intrinsic and extrinsic motivations. Because intrinsic motivation produces greater satisfaction, a gamification strategy should be based on it. However, according to Zichermann and Linder [31], intrinsic motivation is unreliable and variable, so to design a gamification strategy based on it may not be viable or possible. On the other hand, SDT suggests that external rewards, like those used in extrinsic motivation, reduce intrinsic motivations, and therefore, the wish and enjoyment of performing the activity. Zichermann and Linder argued that one strategy is to craft extrinsic motivators such that users feel like or become internalized as intrinsic motivators. Thus, gamification directs people's motivations towards intrinsically motivated, gameful experiences and behavior.

Few frameworks outlining theoretical foundations and how gamification systems can be analyzed exist as reported by Seaborn and Fels [32]. SDT, in particular the concepts of autonomy, competence, and relatedness, was used by Aparicio et al. [33]. They developed a framework divided into four parts: identification of the main objective, identification of the transversal objective, determining game mechanics, and how to evaluate. Blohm and Leimeister [34] developed a framework to estimate how gamification can operate on intrinsic and extrinsic motivators to bring about behavioral change and reframe activities such as learning. A user-centered framework for gamification built upon intrinsic motivation was proposed by Nicholson [35]. This framework is based on Organismic integration theory (a subtheory of SDT), situational relevance, situated motivational affordance, universal design for learning, and user-centered design. Finally, Sakamoto et al. [36] developed a value-based gamification framework for designers aiming to encourage and harness intrinsic motivation. This one is based on TTM that is grounded in Stages of Change and Processes of Change.

Gamification shares many strategies in common with other theories that have been proven to work in the health field, e.g., the cybernetic variations of self-regulation theory. Brian Cugelman studied the relation between gamification and evidence-based health behavior change. Also, Cugelman proposed criteria which developers can use to evaluate when gamification offers a promising framework. He based his study on the assumption that the technology is only persuasive when it employs specific behavior change ingredients. If these ingredients are removed, the technology is no longer persuasive. Cugelman used "ingredient" to refer to the factors that exert persuasive force on people, encouraging them to shift beliefs, attitudes, and actions. He reviewed a number of popular gamification taxonomies to identify the most common strategies used and compared them to a taxonomy of validated interactive behavior change and persuasive design strategies. The author identified seven core ingredients of gamification that have clear linkages to proven behavior change strategies: goal settings, capacity to overcome challenges, providing feedback on performance, reinforcement, compare progress, social connectivity, and fun and playfulness. This study demonstrated that there are some promising links between gamification principles and digital health behavior change science, being no strong link to fun and playfulness in health behavior change approaches. Fig. 7.3 shows the relation between gamification ingredients and validated health behavioral change techniques.

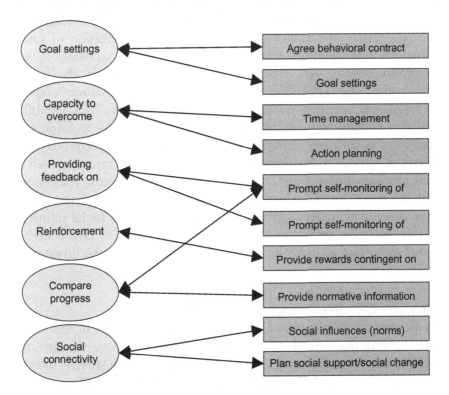

Figure 7.3 Linkages between gamification ingredients and validated health behavioral change techniques.

Payne et al. [37] conducted a systematic literature review about mobile apps used in health behavior interventions. This study described the existing literature on mobile apps, their behavioral features, and evaluated their potential to disseminate health behavior interventions. Payne et al. found 24 studies meeting inclusion criteria, 14 of them involved interventions for physical activity, 4 studies involved diabetes management, 4 for improving mental health, and only 2 studies involved interventions for addiction. All studies involved at least one prominent health behavior theory construct or strategy. Self-monitoring was the most common included in 18 of the studies. Cues to action and feedback were the next most commonly used constructs, followed by social support. Major theories used as frameworks included social cognitive theory and self-determination theory.

Hamari and Koivisto [38] investigated how social influence aids people in continuing and maintaining the beneficial behaviors (physical exercise) promoted with gamification. In particular, they studied how social factors work in parallel to increase willingness to use gamification and continue exercising. Based on the theories of group formation and relatedness, the authors proposed the conceptualization of social influences with three main factors: subjective norms, recognition, and reciprocal benefit. An expanded theorization based on the widely employed models TRA/TPB was used to predict the outcome behavior or behavioral intention related to technology adoption. TPB identifies self-efficacy, attitudes, norms, and sometimes knowledge as important predictors of behavior change. Positive recognition and reciprocity had a positive impact on how much people are willing to exercise as well as their attitudes and willingness to use gamification services. They found that the more friends a user had in the service, the larger the effects were. Also, new understanding on the phenomenon of social influence in technology adoption/use continuance was provided. Other studies about social factors in health behavior may be found in the literature.

7.2.3 The Effectiveness of Gamification

As we previously mentioned, there are several examples in the literature showing that introducing game components improve the intended motivational and engaging values. However, not all studies showed positive effects, and the impact seemed to vary according to the community, users, and product, with some users complaining that gamification was annoying. On the other hand, not all existing studies show significative and determinant conclusions about gamification.

Gamification can be seen as an attractive and effective approach to invoke positive behavioral outcomes. In fact, Hamari et al. conducted a review of peer-reviewed empirical studies on gamification to examine the effects of gamification. They concluded that "gamification provides positive effects and benefits, however, the effects are greatly dependent on the context in which the gamification is being implemented, as well as on the users using it." Behavioral outcomes were found in 21 of the reviewed studies. Only one of those studies was focused on health context. It consisted on a gamified exercise social networking service. In spite of these positive effects and benefits, there is a relative dearth of empirical evidence on its effectiveness. Therefore, further research efforts should be made to obtain empirical evidences on gamification effectiveness. Finally, in order to be considered effective, gamification must sustain its impacts over the long term and offer more than a short-term novelty effect.

7.3 GAMIFICATION AND BEHAVIORAL CHANGE: TECHNIQUES FOR HEALTH SOCIAL MEDIA

As it was already mentioned, gamification is the "the use of game mechanics and experience design to digitally engage and motivate people to achieve their goals" [2]. As a consequence, gamification is used in practice to encourage people to perform a behavior change. Gamification and serious games incorporating gamification ingredients have emerged over the last decade in the health area. Aligned with what Miller et al. [25] and Sergio Jiménez [51] presented, leaderboards, points and levels, challenges, countdowns and virtual rewards are considered as integral components of serious games, which are included even in serious games' design guidelines for seniors [39,40].

A noteworthy portion of gamification studies in the health domain is about physical exercise. Therefore, gamification mechanisms contribute to keeping the users in the "flow zone" which represents the feeling of being complete and energized focus in an activity with a high level of enjoyment and fulfillment towards increased adherence. As it was mentioned above, in these games, players are put in a competitive, collaborative, or hybrid setting to reach exercise goals using gamification such as progress bars, points, poke, badges, bragging, and group quests.

This section focuses on case studies of seniors participating in exergames (serious games promoting physical exercise and activity) with gamification components. The FitForAll exergaming platform as well as a virtual apple picking game were used during pilot trials with seniors.

The FFA platform consists of specifically designed games aiming at elderly exercise and maintenance/advancement of healthy physical status and well-being. It was widely used and evaluated during the trials of the Long Lasting Memories (LLM) project funded by EU [41]. The physical exercises behind the games were in compliance to the ACSM/AHA recommendations [42]. The platform has been widely piloted, and evaluated, with more than 200 participants for about 2 months, exhibiting good efficacy and usability assessment.

Moreover, the virtual apple picking game is part of a suite of exergames aiming at enabling the balance improvement of elderly. It was developed within the Nettrim and GameUp projects [43] using user-centered design at all stages [44]. Socialization, motivational and behavioral factors of the user's were studied upon this exergame [45]. The game was trialed throughout its development process on a bi-weekly basis [46,47] for 2 years with about 15 elderly participants in each meeting.

7.3.1 Daily Achievements Gamification
Given that the primary role of the FitForAll exergaming platform is the delivery of standard physical exercises, it is very difficult to blend most of them with challenging and entertaining game scenarios. As a consequence, the platform's protocol distributes some entertaining aspects within the standard physical exercise. These games are composed of the Apple Tree where the senior collects some apples with his hand's movement, the Mini-Golf where the user's center of mass moves a golf ball to a hole and the Fishing where center of mass is reflected to a boat's position collecting fishes. In all these games, the corresponding score is presented to the seniors alongside the game's remaining time (count down gamification).

A cohort of 15 seniors, playing the games alone, were very excited about their score in all three games. The score promoted the competitive feeling against their previous achievements. The trial facilitator observed that the seniors had their previous high scores in mind comparing it with the current achievement. When technical problems prevented them from achieving a high score, they were asking the facilitator either to exclude their poor performance from the records or to let them try again. Given that most of the seniors have known each other, the competitive feeling was also present and pivotal. The latter was obvious since they kept asking the facilitators about their friends' achievements. Moreover, their scores were a subject of discussion when they met. Worthy of reporting here is a case where a participant requested from the facilitator to play wFFA once more, though she had finished the required sessions, in order to achieve a better score in the golf game than her friend.

Needless to mention that the way the high score is presented to the seniors should be designed in a special taking into account the target group's particularities. When a lady missed to put a ball in the hole, a pop-up message informed her that she lost a life (game lives). The lady complained to the facilitator about the "macabre": way of informing her about her failure by saying that is not proper to tell someone that he lost his "life".

7.3.2 Achievements Award
After the pilot trials, in which 15 seniors participated in exergames, a special ceremony took place. After interacting with each other and discussing about the games, the seniors were given a printout achievements' award including their higher daily score as well as their mean daily and total score in each of the three games. The award had also four photos of their moments during the games (Fig. 7.4).

The facilitators, who were observing the seniors' reaction when they were given the awards, mentioned that almost none of the participants looked at the personal awards. Instead, they hid their awards and asked the others about their achievements. The second reaction of the majority of them was to take a look at their awards, identify the score they went well and announcing this score to the others. They seemed very satisfied as soon as they found out that they had the higher score in any of the scores and tried to find excuses justifying their lower scores. Finally, some participants that were aware of their low scores during the trials did not reveal their scores at all.

Figure 7.4 Achievement award containing daily, mean, and high scores within the pilot trials.

7.3.3 Achievements and Gamification Within Live Social Networks

It is interesting, however, to study gamification effectiveness in live social networks. In other words, how gamification mechanisms work when seniors, interacting with serious games, are in the same place at the same time, forming gaming groups. There are studies proving that gamification, when takes place in live social networks, is more effective [48]. During the Long Lasting Memories (LLM) trials, the intervention was organized in groups of elderly. A number of groups, with different number of participants per group, were created to facilitate the trials management. The groups' size ranged from 1 to 12 users while the number of groups was 46.

Fig. 7.5 shows the dropout level compared to the group's size. The dropout for groups with more than 7 members is significantly smaller than when the group is composed of less than 5 seniors. The dropout rate was 100% when the group was composed of merely one participant [48]. The trials facilitators' observations concluded that contrary to the situation in which seniors are alone when interacting with the games, in larger groups seniors tend to compete or cooperate with higher valence. Playing the same game, side by side, the seniors were giving their best efforts to go better than the other participants.

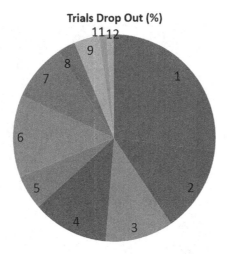

Trials Drop Out (%)

Figure 7.5 Dropout proportion according to the group's size (1–12). There were no group with 10 members.

In the case of a married couple, the husband was complaining about technical problems when his wife had better scores. He was insisting that he was better than her in the game event though the facilitators were calming him down by saying that the scope of the games is not the score per se but the participation.

Obviously, the fact that the participants were in the same place during gameplay, amplified the effectiveness of gamification, which is reflected in the relatively high adherence of 82% for a daily 2 months pilot schedule [39].

7.3.4 Importance of Design in User Behavior

Similarly with the Fit4All platform the virtual apple picking game developed taking into basic consideration the acceptability and usability for the elderly. Central to its development was the users' opinion and the users' observation. The presentation of daily achievements or achievements award is of high importance due to the physical barriers which the elderly may have (reduced vision, hearing problems, etc.). Provision of as little information as possible each time is definitely a key characteristic of exergames for seniors.

It is worth mentioning the example of the virtual apple picking game in which the users were required to catch as many apples falling from a tree and placing them to the correct basket in a specific time as possibly

Figure 7.6 The virtual apple picking game.

achievable. In the center of the screen, the time was counting up to 60 seconds where the game was ending. On the left side the number of apples caught are mentioned while on the right side the apples missed (Fig. 7.6). Despite the fact that the elderly were able to read the font of the letters in this size without effort, during the gameplay almost none of the elderly was following any of the three indicators, resulting to a surprise when the game ended and a question on how many apples they caught in order to compare themselves with the others.

7.4 CONCLUSION

According to the literature review, the case studies and the facilitator's observations, it is implied that games and gamification have an effect on health and it should be considered as essential component of the serious games in the health domain. Gamification, when employed in group of people with face-to-face interaction (even not during the gameplay), encourages the human's intrinsic need for competition. Aiming at engaging user to adhere to a behavior, serious games succeed in increase adherence by being challenging to the user be means of competition or cooperation with friends. However, as it was mentioned earlier in this chapter, further studies measuring latent psychological variables should be conducted to attain more accurate linkages between game mechanics, psychological effects, and behavioral outcomes.

Some studies state that gamification has only short-term benefits without long-term effects [26,49,50]. Even if this is true, gamification remains a very important factor towards engaging user for a short but adequate period until the effects of the serious game are observed by the user. For instance, a senior observes the effects of the physical exercise on his body after a number of exergaming sessions and continues the serious games due to the usefulness factor (according to the TAM—Technology Acceptance Model). Gamification may be essential for the senior to participate in these initial number of sessions.

A motivational framework was proposed [45] based on the input of the users proposed different outputs to increase users' engagements targeting behavioral change on exergames (Fig. 7.7).

In realizing some more implications and take-home messages, one could make the following statements as a crystallization of the experience accumulation from the aforementioned case studies:

• Gamification is an important and integral component of serious games and health-related activities but the delivery means appropriateness is equally important.
• Although seniors enjoy playing against their own scores, they find it more attractive to compete against other users and especially friends as they feel more familiar with them.

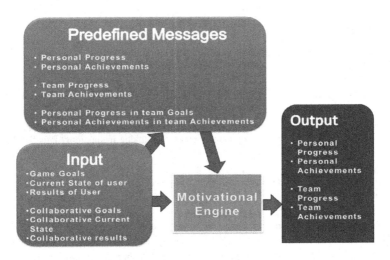

Figure 7.7 Abstract representation of the motivational framework proposed by Konstantinidis et al. [45].

- Gamification effectiveness can be amplified when users are concurrently in the same place, at the same time, interacting with the same gamified activity.

Last but not least, Konstantinidis et al. [45] have also identified a list of persuasive techniques for online exergames in order to foster health social media gamification and behavioral change. These are summarized below as means of take-home messages:

- Display information to encourage people to be more active
- Record and display the user's past behavior
- Use positive reinforcements to improve behaviors
- Make an attractive and friendly user interface
- Provide information at opportune moments
- Use social influence
- Personalization. The goals set for each user are personalized according to her abilities.

Within the limitation of the aforementioned framework, the input of the experts is the most crucial. Positive messages should be tested and adjust to the seniors need in each different environment and game. In addition, socialization has been proven to be an important factor of everyday life, especially for seniors as proved in the case studies of this chapter. To this extent, more efforts on the inclusion of this aspect into exergames should be attempted.

REFERENCES

[1] Greitemeyer T, Mügge DO. Video games do affect social outcomes: a meta-analytic review of the effects of violent and prosocial video game play. Pers Soc Psychol Bull 2014;40 (5):578–89.

[2] Ybarra ML, Boyd D. Can clans protect adolescent players of massively multiplayer online games from violent behaviors? Int J Public Health 2015;60(2):267–76.

[3] Fischer P, Greitemeyer T, Morton T, Kastenmüller A, Postmes T, Frey D, et al. The racing-game effect: why do a video racing game increase risk-taking inclinations? Pers Soc Psychol Bull 2009;35(10):1395–409.

[4] Lanningham-Foster L, Jensen TB, Foster RC, Redmond AB, Walker BA, Heinz D, et al. Energy expenditure of sedentary screen time compared with active screen time for children. Pediatrics 2006;118(6):e1831–5.

[5] Gentile DA, Lynch PJ, Linder JR, Walsh DA. The effects of violent video game habits on adolescent hostility, aggressive behaviors, and school performance. J Adolesc 2004;27 (1):5–22.

[6] Primack B, Carroll M, McNamara M, Klem M, King B, Rich M, et al. Role of video games in improving health-related outcomes: a systematic review. Am J Prev Med 2012;42(6):630–8.

[7] Desmet A, Shegog R, Van Ryckeghem D, Crombez G, De Bourdeaudhuij I. A Systematic review and meta-analysis of interventions for sexual health promotion involving serious digital games. Games for Health J 2015;4(2):78−90.

[8] Gabarron E, Schopf T, Serrano JA, Fernandez-Luque L, Dorronzoro E. Gamification strategy on prevention of STDs for youth. Studies in Health Technology and Informatics 2013;192:1066. Retrieved from http://www.ncbi.nlm.nih.gov/pubmed/23920840.

[9] Gabarron E, Serrano JA, Schopf T, Fernandez-Luque L, Wynn R. Play as a prevention strategy: using a web app to teach youth about STDs. Stud Health Technol Inform 2014;205:1187.

[10] Gabarron E, Serrano JA, Wynn R, Armayones M. Avatars using computer/smartphone mediated communication and social networking in prevention of sexually transmitted diseases among North-Norwegian youngsters. BMC Med Inform Decis Mak 2012;12:120. Available from: http://doi.org/10.1186/1472-6947-12-120.

[11] Li TMH, Chau M, Wong PWC, Lai ESY, Yip PSF. Evaluation of a Web-based social network electronic game in enhancing mental health literacy for young people. J Med Internet Res 2013;15(5):e80. Available from: http://doi.org/10.2196/jmir.2316.

[12] DeSmet A, Van Ryckeghem D, Compernolle S, Baranowski T, Thompson D, Crombez G, et al. A meta-analysis of serious digital games for healthy lifestyle promotion. Prev Med 2014;69:95−107. Available from: http://doi.org/10.1016/j.ypmed.2014.08.026.

[13] Hall AK, Mercado R, Anderson-Lewis C, Darville G, Bernhardt JM. How to design tobacco prevention and control games for youth and adolescents: a qualitative analysis of expert interviews. Games for Health J 2015;4(6):488−93. Available from: http://doi.org/10.1089/g4h.2015.0013.

[14] Staiano AE, Abraham AA, Calvert SL. Motivating effects of cooperative exergame play for overweight and obese adolescents. J Diabetes Sci Technol 2012;6(4):812−19 Retrieved from http://www.pubmedcentral.nih.gov/articlerender.fcgi?artid=3440152&tool=pmcentrez&rendertype=abstract.

[15] Feltz DL, Irwin B, Kerr N. Two-player partnered exergame for obesity prevention: using discrepancy in players' abilities strategy to motivate physical activity. J Diabetes Sci Technol 2012;6(4):820−7.

[16] Lieberman DA. Video games for diabetes self-management: examples and design strategies. J Diabetes Sci Technol 2012;6(4):802−6 Retrieved from http://www.pubmedcentral.nih.gov/articlerender.fcgi?artid=3440150&tool=pmcentrez&rendertype=abstract.

[17] Bul KCM, Franken IHA, Van der Oord S, Kato PM, Danckaerts M, Vreeke LJ, et al. Development and user satisfaction of "Plan-It Commander," a serious game for children with ADHD. Games for Health J 2015;4(6):502−12. Available from: http://doi.org/10.1089/g4h.2015.0021.

[18] Robert PH, König A, Amieva H, Andrieu S, Bremond F, Bullock R, et al. Recommendations for the use of serious games in people with Alzheimer's disease, related disorders and frailty. Front Aging Neurosci 2014;6:54. Available from: http://doi.org/10.3389/fnagi.2014.00054.

[19] Ventola CL. Social media and health care professionals: benefits, risks, and best practices. Pharm Therap 2014;39(7):491−520.

[20] Deterding S, Dixon D, Khaled R, Nacke L. From game design elements to gamefulness: defining gamification. Proceedings of the 15th international academic mindtrek conference: envisioning future media environments. Tampere, Finland: ACM Press; 2011. p. 9−15; New York, NY. Available from: http://blogs.gartner.com/brian_burke/2014/04/04/gartner-redefines-gamification/.

[21] Burke B. Gartner redefines gamification. Gartner blog network. Available from: <http://www.webcitation.org/6fXe9UNwh; 2014> [accessed 24.02.15].

[22] Hamari J, Koivisto J. Social motivations to use gamification: an empirical study of gamifying exercise. In: Proceedings of the European conference on information systems, Utrecht, the Netherlands; 2013.

[23] Huotari K, Hamari J. Defining gamification: a service marketing perspective. Proceedings of the 16th International Academic MindTrek Conference. Tampere, Finland: ACM; 2012. p. 17–22.

[24] Robson K, Plangger K, Kietzmann JH, McCarthy I, Pitt L. Is it all a game? Understanding the principles of gamification. Bus Horiz 2015;58(4):411–20. Available from: http://dx.doi. org/10.1016/j.bushor.2015.03.006.

[25] Miller AS, Cafazzo JA, Seto E. A game plan: gamification design principles in mHealth applications for chronic disease management. Health Informatics J 2014; 1460458214537511.

[26] Hamari J. Transforming homo economicus into homo ludens: a field experiment on gamification in a utilitarian peer-to-peer trading service. Electron Commerce Res Appl 2013;12 (4):236–45. Available from: http://dx.doi.org/10.1016/j.elerap.2013.01.004.

[27] Chen Y, Pu P. HealthyTogether: exploring social incentives for mobile fitness applications. In: Proceedings of Chinese CHI'14, Toronto, ON, Canada, April 26–27; 2014. p. 25–34.

[28] Allam A, Kostova Z, Nakamoto K, Schulz PJ. The effect of social support features and gamification on a web based intervention for rheumatoid arthritis patients: randomized controlled trial. J Med Internet Res 2015;17(1):e14.

[29] Marczewski A. User types. In even Ninja Monkeys like to play: gamification, game thinking and motivational design. 1st ed. CreateSpace Independent Publishing Platform; 2015. p. 65–80.

[30] Hamari J, Koivisto J, Sarsa H. Does gamification work?—a literature review of empirical studies on gamification. In: Proceedings of the Annual Hawaii International Conference on system sciences, January 2014. p. 3025–34. Available from: http://dx.doi.org/10.1109/HICSS.2014.377.

[31] Zichermann G, Linder J. Game-based marketing: inspire customer loyalty through rewards, challenges, and contests. Hoboken, NJ: Wiley; 2010.

[32] Seaborn K, Fels DI. Gamification in theory and action: a survey. Int J Hum Comput Stud 2015;74:14–31. Available from: http://dx.doi.org/10.1016/j.ijhcs.2014.09.006.

[33] Aparicio AF, Vela FLG, Sánchez JLG, Montes JLI. Analysis and application of gamification. Proceedings of the 13th International Conference on interacción persona-ordenador. Presented at INTERACCION'12. Elche, Spain: ACM; 2012. p. 17.

[34] Blohm I, Leimeister JM. Gamification: design of IT-based enhancing services for motivational support and behavioral change. Bus. Inf. Syst. Eng 2013;5:275–8. Available from: http://dx.doi.org/10.1007/s12599-013-0273-5.

[35] Nicholson S. A user-centered theoretical framework for meaningful gamification. Games + Learning + Society 2012;1–7. Available from: http://dx.doi.org/10.1007/978-3-319-10208-5_1.

[36] Sakamoto M, Nakajima T, Alexandrova T. Value-based design for gamifying daily activities. In: Errlich M, Malaka R, Masuch M, editors. Entertainment computing—ICEC 2012, lecture notes in computer science. New York, NY: Springer; 2012. p. 421–4.

[37] Payne HE, Lister C, West JH, Bernhardt JM. Behavioral functionality of mobile apps in health interventions: a systematic review of the literature. JMIR mHealth & uHealth 2015;3 (1):e20. Available from: http://dx.doi.org/10.2196/mhealth.3335.

[38] Hamari J, Koivisto J. "Working out for likes": an empirical study on social influence in exercise gamification. Comput Human Behav 2015;50:333–47. Available from: http://dx. doi.org/10.1016/j.chb.2015.04.018.

[39] Konstantinidis EI, Billis AS, Mouzakidis C, Zilidou V, Antoniou PE, Bamidis PD. Design, implementation and wide pilot deployment of FitForAll: an easy to use exergaming platform improving physical fitness and life quality of senior citizens. IEEE J. Biomed. Health Inform 2014.

[40] Konstantinidis EI, Antoniou PE, Bamidis PD. Exergames for assessment in active and healthy aging: emerging trends and potentialities. In: International conference on information and communication technologies for ageing well and e-health; 2015.

[41] Bamidis PD, Fissler P, Papageorgiou SG, Zilidou V, Konstantinidis EI, Billis AS, et al. Gains in cognition through combined cognitive and physical training: the role of training dosage and severity of neurocognitive disorder. Front Aging Neurosci. 2015. Available from: http://dx.doi.org/10.3389/fnagi.2015.00152.

[42] Nelson ME, Rejeski WJ, Blair SN, Duncan PW, Judge JO, King AC, et al. Physical activity and public health in older adults: recommendation from the American College of Sports Medicine and the American Heart Association. Circulation 2007;116:1094−105. Available from: http://dx.doi.org/10.1161/CIRCULATIONAHA.107.185650.

[43] Konstantinidis ST, Fernández-Luque L, Brox E, Kummervold PE. Seniors Exergaming 2.0: the role of social and motivational aspects on games for elderly's physical training through web 2.0 techniques. In: Medicine 2.0 conference, Malaga, Spain; 2014.

[44] Brox E, Luque LF, Evertsen G, Hernandez JEG. Exergames for elderly: social exergames to persuade seniors to increase physical activity. 2011 5th international conference on pervasive computing technologies for healthcare (PervasiveHealth) and workshops. Dublin: IEEE; 2011. p. 546−9.

[45] Konstantinidis ST, Brox E, Kummervold PE, Hallberg J, Evertsen G, Hirche J. Online social exergames for seniors: a pillar of gamification for clinical practice. In: Novák D, Tulu B, Brendryen H, editors. Handbook of research on holistic perspectives in gamification for clinical practice. IGI Global; 2016. p. 245−76.

[46] Brox E, Evertsen G, Horst-Fraile S, Browne J, Åsheim-Olsen H. Experience with a 3D Kinect exergame for elderly. In 8th international conference on health informatics— Healthinf'15; 2015.

[47] Evertsen G, Brox E. Acceptance of a targeted exergame program by elderly. In: SHI'15, proceedings of the 13th Scandinavien conference on health informatics; 2015. p. 12−8.

[48] Konstantinidis EI, Billis A, Grigoriadou E, Sidiropoulos S, Fasnaki S, Bamidis PD. Affective computing on elderly physical and cognitive training within live social networks. Lecture Notes in Computer Science (including subseries Lecture Notes in Artificial Intelligence and Lecture Notes in Bioinformatics), 7297 LNCS; 2012. p. 339−44.

[49] Farzan R, DiMicco JM, Millen DR, Brownholtz B, Geyer W, Dugan C. When the experiment is over: deploying an incentive system to all the users. In: Symposium on persuasive technology; 2008.

[50] Farzan R, DiMicco JM, Millen DR, Brownholtz B, Geyer W, Dugan C. "Results from deploying a participation incentive mechanism within the enterprise". Proceedings of the twenty-sixth annual SIGCHI conference on human factors in computing systems, April 5−10, 2008. Florence, Italy: ACM; 2008. p. 563−72.

[51] GameOn Lab. Available from: http://www.gameonlab.com/toolkit/; 2015.

CONCLUSIONS

S. Syed-Abdul[1], E. Gabarron[2,3], and A.Y.S. Lau[4]
[1]Taipei Medical University, Taipei, Taiwan [2]University Hospital of North Norway, Tromsø, Norway
[3]The Arctic University of Norway, Tromsø, Norway [4]Macquarie University, Sydney, NSW, Australia

Throughout this book, we have discussed the enormous potential of social media for aiding healthcare and the society in general. Some of these possibilities are already happening, whereas some others are just starting to get explored, and we are sure that the new health-related challenges and the social media solutions are yet to come.

- *For the population and patient empowerment*: The social media has the power to support people with short-term health needs as well as long-term (chronic) conditions. And responsive healthcare providers see the value of engaging, collaborating with getting connected with patients through social media, in order to foster self-management support to achieve positive health outcomes. More details on how social media can support patient empowerment and self-management are reported in Chapter 2, Patient Empowerment Through Social Media.
- *Hospitals and healthcare organizations*: Generally, there are benefits of engaging patients and the public through social media, which point to arguments favoring its use, also in healthcare settings. However, health authorities need more guidance in learning the "ins and outs" of social media for patient and citizen engagement. And this will help to secure more trust in social media use so that we can reap the benefits of posting, gathering, and exchanging information about various aspects of health and care in these everyday interactions (see chapter: Use of Social Media by Hospitals and Health Authorities).
- *Health crisis communication during epidemics*: Researchers have already started to develop methods and strategies for using digital epidemiology to support infectious disease monitoring and surveillance or understand attitudes and concerns about infectious diseases.

A summary of tools and techniques can be found in Chapter 4, Social Media and Health Communication Crisis During Epidemics.

- *For research*: The analysis of the big data that can be extracted from Social Media platforms and other resources (i.e., drug and toxicology databases, "omics" data, electronic health records, etc.) has the potential to improve people's health by means of contributing to a more personalized medicine, offering better health services and contributing in the reduction of costs (see chapter: Big Data Generated Through Social Media).
- *For health behavior change*: Interventions through social media (including primary prevention of chronic diseases; chronic diseases management; and treatment of mental health problems) have proved their effectiveness in changing health behavior−related outcomes, boosting encouragement for future research in this area (see chapter: Social Media and Health Behavior Change).
- *Gamification*: Gamification is strongly related to social media, and game environments are considered as a social media sites. The pros and cons of gamification for health, as well as the persuasive techniques for online exergames in order to foster health social media gamification and behavioral change are identified in Chapter 7, Gamification & Behavioral Change: Techniques for Health Social Media.

Printed in the United States
By Bookmasters